Mennonite Alternative Service in Russia

Mennonite Alternative Service in Russia

The Story of Abram Dück and His Colleagues 1911-1917

Lawrence Klippenstein
Jacob Dick

Published by
Pandora Press

National Library of Canada Cataloguing in Publication

Klippenstein, Lawrence, 1930-
Dick, Jacob, 1931-
Mennonite alternative service in Russia : the story of Abram Dück and his colleagues, 1911-1917 / Lawrence Klippenstein, Jacob Dick.

ISBN-13: 978-1894710213

1. Dück, Abraham. 2. World War, 1914-1919—Conscientious objectors—Russia. 3. World War, 1914-1919—Religious aspects—Mennonites. 4. Labor camps—Russia—History—20th century. 5. Forests and forestry—Russia—History—20th century. 6. World War, 1914-1919—Russia—Medical care. I. Dick, Jacob II. Title.

BX8119.R8K56 2002 261.8'73'092 C2002-902858-2

MENNONITE ALTERNATIVE SERVICE IN RUSSIA:
THE STORY OF ABRAM DÜCK AND HIS COLLEAGUES, 1911-1917

ISBN-13: 978-1894710213

Book layout by Cliff Snyder

The cover photo shows the railway station in Baku, a great industrial city on the Caspian Sea, frequented by the medical service trains.

08 07 06 05 04 03 02 10 9 8 7 6 5 4 3 2 1

Table of Contents

Preface

This work was undertaken with minimal expectations. I was, however, surprised by the many people who showed interest as we talked about it. Encouragement for its completion came from many sides.

So how did it all begin? In 1977 Henry Peters, son of Nikolai Peters, and I decided to meet at a Mennonite conference in St. Catharines, Ontario, to look at each other's family photographs. Our fathers had been colleagues and good friends, so we wanted to compare photographs for similarities. At the conference, a large number of people gathered around us to share in the viewing.

One of these people was Lawrence Klippenstein, then director of the Mennonite Heritage Centre in Winnipeg. He mentioned that the Centre would certainly welcome a collection of such photographs in its archives. Alternative service in Russia, the setting for many of the photographs, was of special interest to the Centre, he said. He went on to say that a written narrative explaining the alternative service photographs would be a very helpful addition to the collection. Since my father had written very little, this project went no further at the time.

After my retirement in 1997, I contacted Lawrence again, reminded him of our discussion some years before. I was again encouraged to continue the project, and given some help to round out the manuscript I had prepared.

I then re-established contact with the Nikolai Peters family, and very shortly thereafter, both of his sons, Henry and Nick, passed away. However, in going through my father's papers again, I found a lengthy memoir by John Mathies, sent to my father by the author.

The greater part of this memoir recalled the time they had spent together in the forestry service and medical corps during World War I in Russia. I knew they were both from the same place of birth and were also recruited for service in the same year. I recalled the frequent visits of John to our home, and how he and my father talked and laughed together for hours on end. John Mathies' memoirs became a further stimulus for my work. I have translated and incorporated large parts of his first-person account in the text that follows.

Many people were involved in shaping the book. I especially want to thank all the families who provided pictorial information for this volume. Among these were the descendants of John Mathies, Nikolai Peters, Jacob Tiessen, Peter J. Sawatzky, and Peter Warkentin. There are also pictures of Abram, my father. Many of the photographs, such as those of Peter Sawatzky, who had worked in a vineyard insect control unit in WWI, had valuable information written on the back. We very much appreciated the help of Albert Willms of Hamilton, Ontario in translating photo notes from Russian into English. The photograph note translations are printed in the shaded boxes (see p. 45, for example).

We also need to acknowledge a number of publications for the use of their illustrations. The photographs of *Als Ihre Zeit Erfuellet War,* edited by Walter Quiring and Helen Bartel as well as those from *Damit es nicht vergessen werde,* edited by Gerhard Lohrenz and published by CMBC Publications, were used by permission.

Our warmest gratitude also goes to Arnold Snyder and Cliff Snyder of Pandora Press in Kitchener, Ontario, who took on the project with unexpected enthusiasm, and patiently helped us work through the process of getting everything together and to the press. Lawrence continued to assist with editing and related work through the completion of the project, as well as writing an historical introduction.

I hope this book will be a resource and aid to help keep alive the memories, and indeed the lessons learned, of a formative period of time in the history of the Mennonite community. It was a time when young men, and others around them, stood up faithfully for the principle of pacifism, and sought to express it meaningfully in a witness of service for their time. In that, they serve as an inspiration for us all.

Jacob Dick
Vineland, Ontario
May 15, 2002

MENNONITES AND MILITARY SERVICE IN RUSSIA

By Lawrence Klippenstein

In the eighteenth century, Catherine the Great welcomed foreign settlers of various nationalities to New Russia. Liberal grants of privileges included freedom of worship and exemption from military obligations. This was important especially for Prussian Mennonites, who established their first settlements in the lower Dnieper area in 1789 (Chortitza) and 1804 (Molotschna). The refusal to take up arms had been a part of Mennonite faith and practice ever since the movement began in Switzerland and Holland two and a half centuries earlier.

The Early Mennonite Involvement With War in New Russia

The issue of state service first became a question of open debate for the new Mennonite colonists when Napoleon marched against Russia in 1812. All the new settlers of the region were urged to support the war efforts as they were able. Those exempted from active service, for example, the foreign colonists, were asked for donations of money. Local officials, including those newly appointed in Mennonite communities, were responsible for setting up collection procedures. Some Mennonite leaders, like Klaas Reimer, objected, stating that such support would be a breach of their non-resistant faith. Small amounts of money were gathered nevertheless, and forwarded to the state treasury.

The Crimean War of 1854-56 brought the matter of public service involvement much closer to home. At the beginning of his reign, Tsar Nicolas I had affirmed his support of the privilege of military exemption, but alternate kinds of service could not be avoided. All the Mennonites, and the inhabitants of other nearby villages, were required to provide wagons and drivers to transport supplies and weapons to the front line, and bring back wounded soldiers to hospitals set up in the colonies. Sums of money, now much larger than during the Napoleonic invasion, were again collected and sent to government offices in support of the Russian war effort.

Some Mennonites lost their lives during the wagon caravanning (*Podwods* in Russian) due to accidents and disease. But Mennonite community leaders received various medals and commendations from the tsar and other government officials, recognizing their efforts to support their homeland in the national struggle of the time.

Records of protests against these alternative forms of state service have not been found to date. One can assume that such voices did exist, and did make themselves heard. Others would object later. For other Mennonites, this experience heightened feelings of loyalty and devotion to the motherland.

The New Military Conscription Laws

Within a decade and a half, the Crimean War would bring the service question back to the Mennonites in another form. From the defeats that Russia experienced in the Crimean conflict, Russian authorities concluded that the armed forces would need to be sharply upgraded. Only then would it be possible to deal with external threats, such as those from Prussia, with strength and victory.

The areas of national well being which came up for consideration in a series of reforms, brought in by Alexander II after 1855, included the army. The tsar energetically supported the efforts of the new Minister of War, Dmitri Miliutin, to bring in new laws to improve the military forces of the country. The first public announcements that such a program was in progress were made in 1870. This publicity informed the Russian populace, among other things, that a form of universal military conscription was to become law, and this law would extend to all citizens of the nation.

The South Russian Mennonites very quickly became aware of these proposed changes, and were unanimously alarmed at the possible conscription of their young men, and about the effect this new law would have on the Mennonites of Russia. No one could be assured of exemption from the extended military service obligations specified by the new legislation.

A contact in St. Petersburg, Pastor Hans, advised the Mennonites to bring their concerns to the authorities in St. Petersburg. A series of meetings to discuss the situation, and to set up delegations, were then held in the churches of the colonies. At least ten delegations were sent to St. Petersburg during the next three years. These representatives were commissioned by their home congregations to contact leading government officials, including the tsar himself, to remind them all of the exemption privileges which Mennonites were given "for all time" when they came to New Russia nearly a century before.[1] The representatives emphasized their faith's conviction that to take human life was against their consciences, and forbidden by biblical teachings, as they understood them.

The delegations tried to do as they were instructed. From the outset, they were assured that the authorities were sympathetic to their concerns, but that exemption from all forms of state service, as had been the case earlier, would no longer be possible. Before very long, they were offered the opportunity to perform non-combatant duties in the form of medical service. But the delegates stated firmly that even this would not be acceptable to their people back home.

Even as these groups were undertaking more trips, another option was being explored. People like Cornelius Janzen of Berdiansk, and some of the ministers like Leonard Sudermann, also from that city, almost at once expressed the view that this development meant emigrating to another country. Janzen openly declared his strong reservations about the possibility of receiving needed government consideration. Very soon, he and his supporters began to promote emigration as the solution to this dilemma.

Before long, contacts were developed with Mennonite of the United States and Canada. By 1873 a delegation of eleven persons, and one other self-supported traveller, undertook a voyage to North America to find out what settlement possibilities existed there. They brought back quite favourable reports from both countries. The delegates of the Bergthal settlement, and the so-called Kleinegemeinde (small congregation) church, recommended Canada to their home congregations. Congregations from the Fuerstenland settlement, whom they had represented as well, got the same message. The rest of the delegates encouraged their congregations to choose the United States. The first families left for the United States that same year, without waiting for the delegates to return.

The Todleben Compromise: Forestry Camps

The new legislation promulgating universal military service in Russia came into force on January 1, 1874. The Russian authorities realized by then that many Mennonite families were getting ready to leave the country. This meant that a significant part of the agricultural population, a very productive sector at that, might be lost. So it was decided that a special emissary of the tsar, Eduard von Todleben, would be sent to the Mennonite settlements to get a closer look at what was actually going on and then send a report to the tsar. He was also authorized to suggest a compromise if necessary, so that some of the people considering emigration might perhaps be persuaded to stay.

On April 16, 1874, Todleben called an important initial meeting at Halbstadt in the Molotschna colony, to address the Mennonite representatives who were invited to attend. His involvements in the Crimean War had given him earlier contacts with Mennonites who had brought provisions to the army there. As a Lutheran, he spoke German fluently and could preach good sermons as needed. He was politely received as a personal representative of the tsar, and was soon able to comprehend the situation.

Very quickly, conversations with some of the leaders convinced him that a good many minds were made up, so that nothing he might say would change their views. He tried to respect all those who simply wanted to know if he would help them get the needed papers and assist them in making the move immediately. However, he also sensed that some people might stay if an acceptable solution to the problem were offered.

Then Todleben brought forward a proposal for an alternate form of service, which the audience seemed willing to consider. He offered the Mennonites the possibility of placing their young men in work camps, either in maintenance stations or forestry camps, where they would serve a period equivalent in duration to the conscription requirements of the army. These camps, he suggested, could be located close to the Mennonite settlements and managed by personnel of their own choosing. They would need to undertake the basic expenses of such camp service, but could also appoint chaplains to serve the units, as they deemed appropriate.

Todleben's visit and presentations generated much discussion and debate. The proposed plan of an alternative service was accepted, with the understanding that the fine details of the arrangement would be discussed further. Mennonite representatives made it clear that forestry camp work was preferable. The government was also ready, Todleben added, to continue the emigration privilege for another six years, until 1880. The first recruitment of Mennonite men for the forestry camps was scheduled to take place that year.[2]

There was no ignoring the fact that the issue of state service, now legislated as universal military service, had been a severe shock to the entire Mennonite community. Tensions between community and congregational leaders deepened as the choice between leaving or staying forced some very difficult decisions, both on the leaders and congregational members. Preparations for leaving deepened wounds for some when buyers of property took advantage of the situation by offering prices far below the value of what was being sold. Unkind words were spoken as old animosities were fuelled afresh by the uncertainties looming before them all.

About a third of the Mennonites in south Russia made their way to Canada and the United States. A total of about seventeen thousand men, women and children boarded trains and set sail during the next six years. The remaining two thirds sought to come to terms with a settlement that was not what they originally would have agreed to. But when faced with leaving as the only other option, this new approach to peaceful service appeared now to permit them to keep the faith, and meet the requirements of citizenship in Russia at the same time.

Final Arrangements for Serving in Forestry Camps

Todleben made the future of Mennonite recruits in the new forestry camp service program look as rosy as possible. On one occasion, he told a delegation: "Your ministers will meet with them regularly. If you train your children as well as you have till now, they will not suffer during the short period of their service. They will not have a difficult time, and when they return they will have learned much that will be useful to them."[3]

All the final arrangements to set up the new forestry camps were completed in the fall of 1879. A special Mennonite planning committee had been set up to fine-tune the proposals in such a way that Mennonite concerns about details would be met. It was agreed that about a hundred Mennonite men would be available each year, as required by the lot selection system. A distribution system would assign certain numbers to take care of planting and caring for new forests, that is to do work like weeding and cultivating, while better-educated unit members would be asked to serve as assistant overseers, secretaries, or accountants.

In June, 1880, State Councillor Bark visited the Mennonite colonies to formally introduce the new legislation dealing with forestry service in Russia. Arrangements were further clarified with the stipulation that the Mennonite community would pay for the cost of housing, feeding and clothing the recruits. The men would be placed in six unit locations during the next three years, beginning with two in the Taurien province, followed by two in the Ekaterinoslav region and two more in the Kherson area.

Not long after this, a special mobile unit to fight vineyard insect infestations was also organized. Its service locale was restricted mainly to the Crimean peninsula. These men usually worked only six months of the year (with time off to go home), but lived mostly in tents, and did not usually enjoy all the amenities of the other camps, such as libraries and chaplaincy support.

If the Mennonites felt unable to cover the food clothing and the housing costs, then the men would be placed in smaller existing camp

2

One of the camps was a mobile unit which had the task of ridding vineyards in the Crimea of damaging pests. It was known as the phylloxera detachment, and was active particularly in the spring, summer and autumn of each year. This picture shows some of the men who were active in this unit.

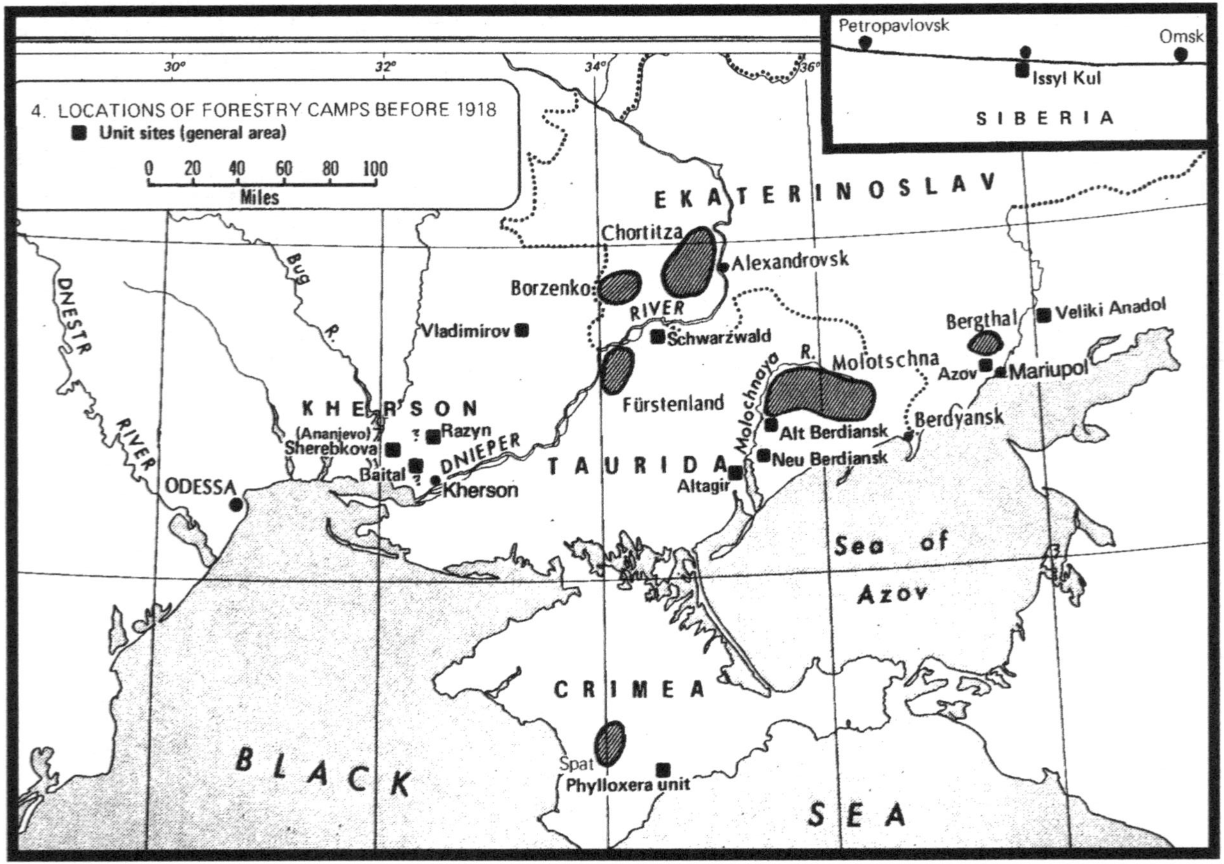

4. LOCATIONS OF FORESTRY CAMPS BEFORE 1918

care locations in outlying parts of the country. This unpleasant possibility may have hastened the decision to accept the government's proposal essentially as offered.

A request to modify the proposals for geographical distribution was granted. This change meant that in the first year, 1880, all new recruits would be placed in the Azov and Anadol forestry stations located not many kilometres north of Mariupol on the Azov Sea, and near the Bergthal colony in the Ekaterinoslav province. Those recruited in 1881 would go to the Vladimir and Ratzyn stations of Kherson, in the vicinity of the Sagradovka settlement, and recruits of 1882 would be assigned to the Alt and Neu Berdiansk camps of the Molotschna colony. Recruits of 1883 would be distributed among all six locations, and in 1884 the rotation would repeat itself.

Men chosen for service were expected to report for duty on March 1 of the year following their summons to service. The regular term of duty would be four years, after which obligated workers could be released into the national reserves. While on duty, the men were required to wear uniforms designed by the Ministry of Forest Domains, and carrying clear designations of rank for both senior and junior administrative officials. Recruits were expected to report for work every

3

Each of the camps set up a small farm to provide food for use in the camps. This is a scene from the Veliki Anadol camp in 1910.

day other than Sundays or holidays, and were paid twenty *kopeks* (about ten cents) per person per day's work completed. The work schedule would involve seven to twelve and a half hours of labour a day, depending on the season. Arrangements could be made for holiday leave between September and March, and visits from family members and friends would be permitted at all times.

Final arrangements stated clearly that there would be no special recruiting centre for Mennonites, but that deferment of service for the first recruits to April 1, 1881, had been granted. With other smaller details worked out, the formalized agreement between the Mennonite community and the government could now be incorporated directly into Russian statute law. There it would stand as Law 914 in Volume II of the third series of the Complete Collection of the Laws of the Russian Empire, abbreviated as PSZ in the Russian title translation, and published under Alexander III in 1886.

Establishing the First Units

The first recruits to be selected participated in a special worship service in Halbstadt on November 14, 1881, and then presented themselves at the recruiting station in the town of Tokmak a few miles away. Lots were drawn by all young men present, and those chosen to serve had their names listed in a special ledger. The process continued smoothly. But there were anti-German mutterings among some local Russians who expressed a strong distaste for these "German privileges."

Service duties at the first camps were to begin on April 1, 1881 at Azov and Anadol. The latter camp had been set up about seventy kilometres north of Mariupol and two kilometres from the Azov Railway. The Khortiza and Molotschna colony included fifty-eight recruits for Azov and fifteen for Anadol.

Jacob Epp, a minister from the Chortitza colony, and long-time resident in the *Judenplan* (Jewish Settlement) to the west, took his place as the first chaplain, alongside Martin Huebert, the first business manager (*Oekonom*) from the Molotschna settlement. By late summer, several men from the Chortitza community had submitted requests for release to join the regular Russian army, rather than work as "slaves" in the forestry units. The churches took the needed steps to subdue this discontent, and to "punish such disrespect for the fathers." Peter, the son of the chaplain, would recall that one heard a great deal at the outset about "the bad behaviour of the young men," but that "great improvements were noted as time went on, thank God."[4]

The second year of camp activity produced extensions of the program, and established a pattern of operations which would continue throughout several decades. An addition of twenty-two men at the Anadol unit brought the contingent there to a total of eighty-six. Nineteen of the initiates (*novobrantsy*) came from the Ekaterinoslav province and three came from the Warsaw area. In mid-April Martin Huebert submitted a report where he shared notes on the work. "We completed the planting of the saplings by Easter. About 270 acres are planted each year. We have now done about 2,700 acres altogether, with 8,100 acres more to go. This forest area was established in 1843 so some trees are a foot and a half in diameter. Seeding is completed for a (local small farm) except for corn, potatoes, flaxseed and watermelons ..."[5]

Work details, on duty at more distant points from the main camp, sometimes set up temporary quarters for shelter, resting, etc.

By this time another camp had been opened at the Vladimirov location in the Sagradovka settlement south-west of Chortitza. A dedication service for the new barracks

These are the camp buildings erected at the Ratzyn forestry camp.

there was held on February 27, 1882. A few days later, several church leaders of the area instituted a special memorial day service in honour of Alexander II whose tragic death had occurred on March 1 of the previous year. About three hundred fifty people gathered for the celebration. By the end of the year yet another camp became operative at Ratzyn. A fifth camp began its work at Neu Berdiansk in 1883, with Alt Berdiansk following quickly to complete the projected series of six in all.

Everyday Life in the Camps

Physical arrangements were not, on the whole, difficult to make. However, camp morale and discipline became issues very early on, as noted above. Reports of laziness, unseemly venturing into nearby Russian villages, insubordination of the men with respect to chaplains and business managers, and other forms of undesirable behaviour would surface from time to time. These concerns would then find their way into the Mennonite press, which consisted largely of North American papers at the time.

The conference of the ministers of the Russian Mennonite colonies tried to deal with these problems to the best of their abilities. Special appointees were sent to discuss unit needs with the men, and sometimes extra meetings were called to look for innovative ways of improving the situation. Singing and instrumental music were introduced as a way to encourage positive interaction. An orchestra was organized at Anadol, but dance music was forbidden. Libraries were also set up, to offer informal adult education.

Forestry camp workers settled into basic routines: it seemed that the arrangements were basically satisfactory as far as outward requirements were concerned. A good deal of useful work related to forest care and new tree growth was being completed at all the camps. Government-appointed officials tried to understand the concerns of

Men at work in the Schwarzwald (Sherbkovo) camp.

their service men, and usually collaborated with the larger Mennonite constituency to deal with problems as they arose.

The meaningfulness of camp work under these conditions was certainly another matter of concern for many of the recruits. Its direct relevance to pacifism and its providing some kind of peacemaker message to the nation, as a tenet of faith, was not always clear. Some thought such connections were not being made at all. In fact, in some people's minds the need to consider other service options had existed right from the start. Before very long, the opportunity to try new approaches would be at hand.

The War with Japan

By the turn of the century, Mennonites of Russia had experienced two decades of state service in forestry camps, as an alternative to taking up arms. It had been a costly venture, and there were frustrations which had not always been easy to overcome. But forestry service, as defined in the arrangements of the 1870s, had become an established institution. It still seemed to be the best solution available at the time.

The war of 1904-05 with Japan gave these arrangements new relevance and a fresh immediacy in the Mennonite community. There was a quick identification of all Russians, including Mennonites, with the plight of the nation so "traitorously" attacked, as one Mennonite writer put it. Perhaps more than ever before, it seemed like a national duty to do something for Russia.[6]

Heinrich Dirks, a Mennonite minister, stated in his report to the 1904 Mennonite yearbook: "Though because of our confession we cannot take part in active military service, we may nevertheless in other ways contend for the good and the benefit of our fatherland. Forestry service, which we render, is after all a useful service, and we are generously supporting relief of suffering caused by the war. Our beloved Monarch will surely not ignore the warm patriotism which is so characteristic of Mennonites."[7]

Responses were indeed quick to come. Regional district leaders of Halbstadt and Gnadenfeld in the Molotschna settlement sought to rally their residents to help in the traditional Mennonite manner. Landowners of the area agreed to contribute fifty *kopeks* (about twenty-five cents) per *dessiatin* (equivalent to 2.7 acres, or 1.092 hectares), and with similar responses in the Chortitza and Yazykova regions, the community raised a total of 83,940 *rubles* (about $42,000) altogether. By June, gifts to the Red Cross were mounting steadily. Requisitioning of horses began in summer. Before long fifteen hundred animals were sent on to government war supply stations for army use.

A reporter for the Odessa *Zeitung* had this to add about what was happening: "Let us remember the army in prayer daily, and generously to relieve the suffering of the wounded, as well as the needs of the widows and children of those killed in battle ... to show our gratitude that so far we may remain at home in peace. ..."[8] Community leaders agreed to recommend strongly that their villages should undertake the care of wounded soldiers, and support of their families for the duration of the war. A voluntary levy would be proposed to create the funding for this initiative.

Four young Mennonite volunteers outfitted in the uniforms of the German Red Cross, which sponsored a mobile hospital unit at the Far Eastern front of the Russo-Japanese war. They include a Mr. Peter Dyck seated at the far right, later an immigrant to Canada.

Front-line service opportunities would provide a new element of involvement as far as Mennonites were concerned. When the volunteers for medical services in the villages proved to be conscientious and hard working, the government began to look for Mennonite stretcher-bearers and servicemen for the Red Cross units at the front.

A group of seven Mennonite volunteers left in early November, 1904, to join a field hospital donated by the German government. Upon arrival in Harbin (China) nearly two months later, they were assigned to a large building and set up two additional barracks which they had brought along. Correspondence with friends at home provided a glimpse of what was happening on the front. Heinrich Friesen, one of the first to go, wrote: "Warm thanks in the name of all suffering soldiers in the far east to Abraham and Kornelius Klassen, Ufa province, at Davlekanovo, and to all others who contributed to the gift of 1131 warm shirts."[9] He also wrote to Elder Isaak Dyck of the Chortitza colony, urging him to plead for more front-line volunteers.

Under Dyck's leadership, the community agreed to press for more recruits, and over-subscribed the local goal of twenty men, when forty-two signed up to go. Nearly half withdrew later, but the remaining twenty-two received a posting to leave in March, 1905. At Ekaterinoslav the men were outfitted with boots and clothing, given the necessary papers, and sent on their way. This group served for three months at a Red Cross hospital at Tungusen. Six of the medics were ordered to return home and report for forestry service in October that same year.

These demands for front-line volunteers created a new and controversial dimension for the home front planners of war support and alternative service. The enthusiasm of the young foresters looking for more meaningful service opportunities was generally not shared by their parents and most of the church leaders. Jacob Hoemsen visited friends while en route to the front, and found some of them less than encouraging. He later wrote: "I also visited my friend H. Martens in Kalinovo. That Mennonites should participate in the war as medics was something quite new. Mrs. Martens voiced her disapproval in unmistakable tones. 'All of you ought to have your seats warmed.' I did not allow it to bother me."[10]

Disagreements about a Mennonite role in the war became a conference issue at a special session convened at Karassan in the Crimea on September 17-18, 1904. In what was hoped might become the official position, Elder Abram Goertz read a paper in which he expressed opposition to volunteering for medical service, as had been done in past months. The Privilegium (Charter) of alternative service of 1874 had been negotiated, he stated, to exclude all forms of service other than forestry work. Alternative service was limited to forestry work for two reasons: it would allow the recruits to remain in groups, and such service had no direct contact with the army.

In Goertz's view, volunteering contradicted these two principles, especially when it happened, as it had in this case, without consent of the church. He asked forestry camp leaders and others to notify church authorities if a movement for volunteering became apparent in their units. Goertz believed the whole matter needed to remain in the hands of the leaders who had originally negotiated forestry service arrangements with government authorities. While not denying that medical service was worthwhile, Goertz preferred to see it happen in Mennonite hospitals.

In fact, the first group of volunteers had included a number of persons from such camps as Vladimirov in the Sagradovka area, who had requested (and easily obtained) a transfer from forestry work to duty on the Far Eastern battlefront. Hoemsen himself was one of about thirty men stationed in the region of Uralga in the village of Nerchinske, where, in his words "all were well thought of by the authorities."[11]

Elder Heinrich Unruh held out, with equal conviction, for a somewhat different position. Utilizing the story of the good Samaritan as a biblical example, Unruh took issue with the view of Goertz and others who resisted the volunteers. Money, said Unruh, was not enough; it was too impersonal and such forms of service alone could too easily look like an excuse to remain comfortable at home. He reminded the group that the Taurida hospital wanted twelve Mennonite medics, and the Ekaterinoslav hospital wanted only Mennonite men.

The war came to a rather quick end, as it turned out, but it did not end favourably for Russia. Mennonites sent a special letter to the tsar expressing "profound gratitude and awe at the wisdom of His Royal Majesty for bringing peace to the nation again."[12] Abram Kroeker, editor of the Mennonite newspaper, *Friedensstimme,* had been depicting Mennonite support activities as "an opportunity to repay with gratitude what one owed to the government and the Fatherland ... and at the same time prove that the accusations frequently heard, that we are anti-Russian, do not reflect the facts."[13]

At least one of the front-line volunteers, Dr. Nikolai Friesen, had died of exhaustion while on duty. Mennonite historian, P. M. Friesen, estimated war support to have cost the Mennonites a total of three hundred thousand *rubles* (about $150,000) in all.[14] There were, however, some who felt that even more difficult times might lie ahead. Freedom of belief and conscience, as important as they were, had a fragile footing in Russia. For minorities, the struggle to retain respect and maintain liberty needed to be enacted in a different way. Militants among the people might not be easily pacified. When the Duma appeared, it would attempt a parliamentary approach to achieve equality and recognition of minority views.

The Camps Under Review

At home in their villages, Mennonites were also reconsidering the experience of alternative service at another level. The Russo-Japanese war had caused some Mennonites to say quite openly that the forestry program should be abolished. Some of these same voices called for alternative military service for all, even if this service included non-combatant forms. Writers to the Mennonite press found support for the view that Mennonites really had no moral right to ask for special privileges, least of all in matters of making peace, since "they had as much quarrelling among themselves as others." Another said plainly: "We should offer medical service now."[15] Mennonites were, by this time, as willing as anyone else to defend themselves with weapons if attacked, and some were even hiring armed help.

In the meantime, Mennonites were asked by the government to open up more forestry camps. A call for two camps coincided with a Mennonite request for concessions, and permission from the government for one camp only. Plans for a new site in Siberia had, in fact, already been drawn up. Issyl Kul, on the Trans-Siberian Railway, midway between Petropavlovsk and Omsk, was soon approved by the government so that the first recruits could be stationed there by late March of 1914. This small group of thirteen men brought the total of young Mennonites in forestry service to just over twelve hundred. That increased by nearly a hundred the total of 1123 who had been on duty the year before. In 1912 there had been 1085 in all.

Rumblings of problems persisted, however. The forestry president, David Claassen, passed on a message during this time, noting that most of the camps were not really meeting government expectations, as far as forestation targets were concerned. Trees planted twenty or thirty years before had not grown to expected maturity, and indeed were sometimes hardly to be found at all. Claassen noted government discussions about closing the camps altogether, or perhaps moving all the camp service men up near the Chinese and the Persian border to add some greenery to the desert.

In the May 1914 meeting, forestry service representatives did not seem unduly alarmed. Few would have guessed that far more stressful times lay just ahead. Most of the business agenda dealt with routine matters, which simply needed formal acknowledgement and consent.

The Balkans, however, had already burst aflame with fighting. Once again, Mennonites refused to take up arms, but in a few weeks Russia would also be at war.

War!

The Mennonite newspaper, *Friedensstimme* (Voice of Peace), ran an issue of its regular publishing schedule on July 19, 1914, when Russia and Germany declared their intent to do battle with each other. For its readers and the entire community the key concern was: What will our obligations be as Mennonites of Russia? How will the long-standing belief in non-resistance, or refusal to take up arms, relate to this situation? Editor Abram Kroeker sought to highlight these anxieties in his own question: "Do we not owe it to our government and Russian neighbours to show that if a war ... broke out, we would be ready to serve the interests of the Fatherland [Russia], and to help the needy?"

There were two things Kroeker felt that Mennonites could do with a clear conscience. One was to make donations to the Red Cross and offer aid in the hospitals and care of wounded in the colonies. This service was reminiscent of the days of the Crimean War. Secondly, they could organize medical units in the colonies and make them available also for the lines up front. To a degree, this echoed Mennonite involvement in the Russo-Japanese War.

Preferably, Kroeker added to his second suggestion, these units should be staffed by young Mennonite volunteers whose motivation would not be adventure, but Christian service.[16] *Der Botschafter* (The Messenger), the other widely-read Mennonite newspaper of the colonies, shared these sentiments. In his editorial, D. H. Epp said: "We need to show that we have kept the promise of faithfulness made to our forefathers ... our confession forbids us to spill blood, but binding wounds we hold to be a sacred duty."[17]

This group of Mennonite nurses served in the city of Simferopol in the Crimea.

The Mennonite colonies felt the impact of impending fighting, even before an initial public declaration. On July 17, a Ladekopp businessman, Peter Dyck, had already written in his personal diary: "In the afternoon came the unexpected news about the mobilization. After lunch, all of us had to take our horses to Halbstadt where twenty-five were requisitioned from four villages ..."[18]

Next day Dyck had to take twenty soldiers to Halbstadt, and wrote that evening: "This morning I went to Tokmak with a bicycle. Crowds of people had gathered. About noon hundreds of soldiers travelled through. What will happen? People are distressed ... I am quite confused ... this is too sudden and unexpected."[19]

In a matter of days, local communities had organized self-styled efforts to render support and aid. Both the Chortitza and Molotschna leaders met on July 22, the third day of the war. Sessions in Chortitza began with prayers for the tsar and the country, and a resolution to send to Nicholas II a pledge of loyalty and an offer of assistance in healing the sick and wounded soldiers. Elsewhere in the nation, the tsar's call to arms brought together just over three million men. Only four percent failed to heed the call to defend Mother Russia.

In the Mennonite community, plans were immediately laid to establish a hundred-bed hospital for recuperating servicemen from the battle lines, men who would presumably soon be transported inland in large numbers. The assembly also agreed to house soldiers in private homes; to make an initial donation to the Red Cross of seventy-five thousand *rubles* (about $37,500); and to gather further donations, both at the district offices and at Nikolaipol, to help with medical services required by the army. Finally, it was also decided to set up a committee of fourteen persons to supervise the granting of aid to families of soldiers, and to make twenty-five thousand *rubles* (about $12,500) immediately available for this purpose.

The Molotschna Church Ministerial suggested that some persons might want to volunteer for hospital service on the front lines. Dr. J. Esau from the Crimea received an appointment as a doctor in a new Red Cross hospital and soon took on other responsibilities with the organization. Another doctor, John Isaak, proposed that Mennonites set up their own field hospital of fifty beds, with two or three doctors, six nurses and fifteen orderlies. The costs, he added, would be no higher than the total money collected by the Mennonite municipalities in the war of 1904-05. He felt that the municipality leaders and elders of the churches should give leadership to such a project and begin at once.

There was an outpouring of offers to help from the Mennonite communities of the country. Mennonites and Russians at Millerovo, in the Don region, also elected a committee to collect money and goods for distribution. A total of four thousand *rubles* (about $2,000), and about twenty-four tons of flour from the local mills, came in almost immediately. A municipality collection at Alexanderthal, in Alt Samara, brought in three thousand *rubles* (about $1,500) to assist the wives of Russian soldiers in three neighbouring villages. At Davlekanovo, the community agreed to organize a twenty-five-bed field hospital, and also pledged support for soldiers' families.

Back in the Camps

In the forest camps, many young servicemen hoped for a quick transfer to the medical service to work as orderlies in hospitals. At Sherebkovo twenty men volunteered immediately, while at the Alt and Neu Berdiansk camps others stood by to leave at any time. Some men took leave from their regular duties to help families of soldiers with the fall harvest. Foresters tended, of course, to discourage departures; however, the weariness of forest routines, the appeals of the Mennonite press, and the desire for more direct involvement in national service gave the recruiters of medical orderlies excellent prospects for success.

When forestry president David Claassen returned from Petrograd at the beginning of August, he was able to answer many recruitment and service questions more clearly. Mennonites were informed that all reservists up to forty-three years of age would be obligated to serve along with the regular age groups. Not even teachers would be exempt.

Men in the camp were quartered in barracks, or dormitories, constructed especially for this purpose. The photo shows the inside of these private quarters at the Alt-Berdiansk forestry camp.

About this time, the Chortitza and Molotschna municipalities received a request from the central *zemstvo* (municipal council) office for four hundred orderlies.

Responses to this request were mixed at first. A Chortitza membership meeting quickly accepted Duma member Herman Bergmann's suggestion that the Mennonites recommend a change of service to include medical work as well. In the Molotschna Ministerial, on the other hand, some felt that the government's wish for a new forestry camp at Issyl Kul indicated satisfaction with existing service arrangements, and hence no change seemed necessary.

By August 5, 1914, the *Botschafter* editor could announce that the needed volunteers were all registered. Prince Georgii Evgenovich Lvov, head of the All Russian Zemstvo Union which was coordinating medical work, could announce a week later that the first group of Mennonite volunteers was to be organized in Moscow immediately. In mid-August the initial contingent of fifty-four men headed for the capital to receive training and assignment to active duty.

Two large volunteer organizations, the All Russian Zemstvo Union (ARZU) and the All Russian Union of Towns (ARUT), formed the basic structural framework within which most Mennonite orderlies and related personnel could be integrated into the total war effort of the country. As part of the elemental patriotic upsurge of the Russian people, these two bodies emerged in the very early weeks of the war to mobilize the general public to provide extra services for the soldiers on front-line duty. Their primary aims included the organization of relief, particularly in the form of aid to sick and wounded soldiers. This was a facet of the war effort which Mennonites generally found quite compatible with their capabilities and beliefs.

The Mennonite Service Centre in Moscow

By the start of World War I, the first steps had been taken in Moscow to establish a Mennonite Unit Service Centre under the auspices of the ARZU and ARUT. At the head office of the former, on Nikitski Boulvar, the first Mennonite volunteers met the top administrators of Russia's major war relief agency. When their first guide, enlisted to find living quarters, began to misappropriate the unit allowances entrusted to him, the Mennonite men insisted that they be directly responsible for any monies available for the maintenance of the units. With these arrangements for purchase of food and clothing, as well as the appointment of a group leader, Armin Lehn, the foundation of the Centre had been essentially established.

One of the men described the details of getting started in a letter to his family. He wrote, "the first men arrived on August 12 ... we were well received and given twenty-five *rubles* (about $12.50) monthly, along with uniforms. ... The Union expects to place fifteen hundred men in trains. ... They shall bring home wounded from the front, and care for them en route to places of recuperation ... we are now 135 men all impatient to be doing the actual work ... cramped together in several rooms like herring in a barrel ... we have our own cook and kitchen and are organized much as at the forest camp ... we find much satisfaction in our work, and the wounded are grateful ... the Society has high expectations."[20]

Nine instructors were assigned to oversee the work in the train stations while two, Franz Thiessen and William Klassen, served as liaison officers to keep up contacts with the main office. The instructors' job was to take new recruits to their work sites, to supervise their activities, and generally offer advice and counsel as needed. A vacant factory surrounded by a high wall became the headquarters building for the Centre. It would be the main point of deployment for five to six thousand Mennonite medical servicemen during the course of the war.

Remembering these experiences, Franz Thiessen described the preparation of the trains, which involved cleaning old freight cars, installing bunk beds, and furnishing them with medical supplies and bedding. He noted that each train carried a staff of thirty-five to fifty persons.

Preparations at Home

Meanwhile, the Mennonite communities were establishing special hospital facilities for wounded soldiers. On August 15, 1914 the Halbstadt municipality agreed to open up forty-five beds for this purpose, thirty in Muntau and fifteen in Ohrloff. Fifteen more were available in private homes. The Gnadenfeld municipality similarly set out to provide thirty beds at Waldheim under the supervision of Dr. Heinrichs. Sheets and other supplies came from nearby villages, and monies were raised to support medical facilities at Alexandrovsk and Melitopol.

Local leaders sought to encourage strong support of these ventures, and chided those who seemed to be shirking their duties. Heinrich Unruh wrote in his announcements to the eligible men of the community: "Let our motto be 'With God for Tsar and the Fatherland'."[21] In a sermon, John Klassen said, "Let us offend no one so that they will have no reason to question, as though we are not true Russian patriots, whose hearts beat warmly for the Fatherland."[22] The president of the forestry services, David Claassen, spoke for the leaders as a whole: "The present situation will reveal in a special way what faithfulness and devotion in the Fatherland each citizen has."[23]

Mennonite reservists received a call to appear in Ekaterinoslav on September 3 and reservists in Siberia were to appear on September

10. One of them described the departure in his area thus: "We went with our parents to Halbstadt. Many teams were there. I have never seen such a crowd at the church yard. ... Then they were off to the train station with hundreds of teams, many automobiles and several hundred people. Eight to ten train cars were ready."[24]

Some men arrived in Ekaterinoslav the day before their appointments there. They received temporary lodging in several large buildings, including a girls' school which held eight hundred men. For their assignments, the recruits could choose either to go to forestry camps, or take up some kind of medical work in trains, hospitals or a related facility. Destinations and points of service were then determined by the authorities, and thereupon the men entrained in small groups to head for various parts of the country to take up their service duties.

The first three thousand men who answered the call to the colours were found fit, for the most part, and quickly sent on their way. By September 7, some had already left, and within a week most of the rest had been assigned their duties and moved on. One participant estimated that of that first contingent, about seventeen hundred were designated for forestry work, and the rest presumably would be heading for medical service pending further assignment details at the Service Centre in Moscow.

Good friends frequently chose to stay together. One group of four hundred found itself entrained for Petrograd with no further information about ultimate destinations. In the end they found themselves in Moscow, where they were given to the two relief organizations through choice by lot, all to serve as medical orderlies on the front lines. The ARZU (All Russian Zemstvo Union, the organization which set up and administered the work of Mennonite medics in WWI) alone had opened up 462 hospitals in August, with about twelve hundred more to follow in September and October. After Petrograd and Moscow, the largest hospital centres were in Kiev, Kazan,

A wounded soldier being placed in a medical train for transport to a hospital away from the front.

Nizhni-Novgorod, Saratov, Voronezh, Tula, Kharkov, and Rostov-on-Don. The first wounded men from the battlefields had already arrived in Moscow on August 6, 1914.

A few weeks later, in late September and early October, the first Mennonite medics left for the Prussian lines on the north-western front. Each medical supply unit had eight cars, completely furnished with food and supplies for the wounded, a kitchen with three cooks, a dispensary, a fourth class car with thirty-three orderlies, a third class car with a doctor, three paramedics, a manager and additional nursing personnel.

The Caucasian Front

Most Mennonites sought to avoid the Caucasian or Turkish front if at all possible. Travel difficulties in rugged terrain, the high risk of diseases such as typhoid fever and malaria, and the fear of the native tribal population made the area an undesirable option for many. However, for those who resided in this general region, and for those from the Urals and Siberia, the Caucasus was actually closer to home, making the extra risks seem less formidable when weighed against the advantages.

Casualties were high at the outset, with the Russian army largely defensive in its objectives. The army did not do well in the early clashes with the Turkish armies of the area. Extreme temperature changes and the high incidence of disease heightened the demands, as well as the health risks of working in field hospitals there. Orderlies frequently shared sick beds with the soldiers, and a number of the orderlies died. Many hoped for a transfer, and the plan, at that point, to form a "Black Sea Unit" gave some the hope of transfer.

Twenty Mennonites from the 7th Medical Corps and the orderly trains of the Caucasus met at Tiflis for preparations. There they learned that they would accompany an attack on Constantinople from Odessa, Sevastopol and Batum. The attack had been designed to assist the British in their assault on the Dardanelles. Sent first to Batum, they could see no signs of ships being loaded, so they were shipped to Odessa, only to learn that the plans for an attack had been cancelled.

Of this situation, one medic observed that "the Goeben and Breslau (German battleships in the Black Sea) would have destroyed the entire Russian fleet if the plan had been carried through."[25] The medical unit was moved to the western front, where the Russian army was in retreat as the Germans were breaking through. Some Mennonites were taken prisoner in the conflict there, while the rest could finally return to Kiev.

Mennonite experiences at Kara Kalissa, and at the hospital at Prochladnaja, involved other typical kinds of work and service during the Russo-Turkish battles of the period. The hospital included many smaller and larger tents which were lined with blankets. The tents were heated by wood stoves. Two orderlies were assigned for every tent and thirty beds. Orderlies worked as watchmen during the night as well. When new patients arrived, mattresses had to be filled with fresh straw, tents cleaned, wounds dressed, and patients transferred to inland hospitals as soon as possible. Many Mennonites became drivers, taking wounded soldiers to Kars. At Trapezund a new hospital was staffed mostly by Mennonite personnel.

Forestry Camps in Wartime

The new opportunities of medical service work did not attract all the Mennonite servicemen of the war years. Indeed, all forestry camps established prior to the outbreak of the war continued their operations, including the newest one established at Issyl Kul in Siberia. Forestry president Claassen had said right from the beginning that both new recruits and reservists would be placed in forests as well as medical service as alternatives to conscription in the armed forces.

News from the forestry groups dispersed at the call-up of September 3, 1914 reached the Mennonite press a few weeks later. The ordered camp life of the past thirty-five years did not always prevail in the new war-time locations. One of the men wrote from the Savranskoe forestry camp, "As soon as we arrived, we were assigned to the forest eight kilometres away to serve as forest guards. We each got about two thousand acres to look after. Our guard cabins are about five kilometres apart so we cannot stay together."[26]

Some men, like those at Tambov, seemed enthusiastic about the early prospects of their work. At first their forester had been uncertain about how to utilize his new aides, but then they too had each received a section of forest to guard. Others were less sanguine about their lot, mentioning their homesickness, and distaste for the local Russian diet of bread with *salo* (salted pork), onions with salt, potatoes and tea. Sometimes they had to endure a shortage of bread.

Heinrich Janz talked about having to take care of gardens. He spoke of over-crowded quarters, and also mentioned that they were not getting the best spiritual food. Since no minister had joined their group, they were using a book of sermons, he said. "Bombard us with letters," he added, "Do not spare ink and paper."[27]

Other Types of Work

In the Vitka-Kostroma region, young Mennonite recruits supervised a large prisoner-of-war camp. Here an initial unit of a hundred Mennonites formed the nucleus of what would ultimately become a group of seven hundred. Most of the labour for clearing and gathering birch logs used to manufacture components of poison gas came from the force of about four thousand Austrian prisoners held in the region. Many of the supervisory positions in the workshops, kitchen, or office went to Mennonite personnel. In time about twenty men were asked to take over guard duties at the prison camp so that they might meditate conflicts which arose between the Russian administration and the prisoners.

Road construction was another dimension of forestry work, especially in the Crimea. About eight hundred men had been placed there at the September 3 registration. Forty men had gone ahead to build barracks as residences, and on September 22,1914 fifty others, mostly volunteers, left for the Crimea. Jacob J. Thiessen wrote about his assignment in the Mount Tschuschelj region: "It was our assignment to build a road from Simferopol through the mountains to Yalta. ... Everything was done by hand. We were sixty men in one barrack house. The building was primitively built, with walls of branches plastered over with clay ... when it rained it often did so longer inside than outside."[28]

The End of Alternative Service Under the Tsars

The abdication of the tsar and the beginning of the Provisional Government allowed the Allies to keep up the pressure for Russia to stay in the war. One positive outcome for Mennonites was that the pressures created by the land appropriation laws in the Duma, under the tsars, were now reduced. The matter would come under review as further immediate expropriations were halted.

At this time, the Mennonite privilege of military exemption became more secure. As government ministers in office changed with increased rapidity, there were short periods when conscription to military service seemed imminent. However, it became clear that the new government intended to maintain the hospital train and forestry alternatives for the duration of the war.

A special Mennonite delegation to Petrograd, seeking to end land expropriation pressures, asked that all forestry workers be transferred to hospital trains immediately. It also proposed that the government undertake the clothing and care of all medical servicemen, that the families of recruited men be supported financially, and that the travelling costs of medical servicemen be taken over by the state. Most of these requests were denied outright.

As expenses for the program rose to about a hundred thousand *rubles* (about $50,000) a month, the financial aspect of alternative service became a crucial issue for the Mennonite communities themselves. It had become much more difficult to collect the self-levied taxes needed to raise these support funds. Many shortages of supplies in the camps were reported, and the wives of the Mennonite servicemen were suffering also.

Administratively, it was agreed that the old system and its forestry president (still David Claassen at this time) would be abolished, to be replaced by an "Executive committee of the Representative Assembly of the Mennonite Churches for the Upkeep of the Forest Units." H. B. Janz took over the position of chairman of this new body.

Units of orderlies at the front were finding themselves increasingly isolated, and left to fend for themselves in the new conditions of the front lines. The change of government had created a general instability. Some men went to recruiting offices to look for dismissal, and others, particularly teachers, simply began to return home. The Mennonite All Union Conference tried to reassure the men that support at home remained strong. Under the Kerensky regime, which took over in July, the All Union Conference attempted to maintain that their traditional views regarding military exemption and alternative service had not changed.

Those who remained on active service in the forestry and medical units faced increasing pressure from soldiers' groups to take their places in the ranks of military personnel. In response to these voices, Mennonite unitmen reiterated their traditional pacifist views in public statements which were duly noted in conference sessions back home. Even so, there were some in the churches who were ready to give room to persons who might not take the older traditional view, and called for toleration of those who might, under changing, more threatening conditions, take up arms in self-defence as a last resort.

A New Regime Takes Over

A complete end to the service program portrayed in this story did not come until the Bolsheviks took over control of the government in November, 1917. The subsequent civil war would test Mennonite principles of non-resistance in a way never experienced before. A self-defence corps (*Selbstschutz*) made up of Mennonite men would be formed before those turbulent months came to a close and the new Red regime firmly took control of the country. The corps always claimed that its intent was to provide protection against roving Makhno bands of southern Russia, and that with the reassertion of regular government controls, Red or White, these arms would be laid down. These claims proved true.

Most of the Mennonite men had returned home by mid-summer of 1918, having done so either at the time of the Revolution itself, or when a peace treaty signed that year at Brest brought the war with Germany to a formal conclusion. Some had gone on leave during those months and simply never returned.

Isolated units of forest guards remained on duty formally in northern Russia for a time. A reporter noted that the Executive Committee of Representatives for Forestry Affairs still needed to raise support for five thousand men during the final months of 1917, that it now had to repay the loans taken out for those payments, and that its obligations to Mennonite men still in service at the time would continue into the new year.

It would be mid-summer of 1918 before all the forestry camps were dismantled. Some of that property had indeed been confiscated by the Red government, and in some cases camps were raided by neighbours and much property disappeared. Despite the appropriation and pillaging a good deal of the land was sold. Some allowances did remain unpaid as last-minute appeals to send in dues fell on deaf ears. A Mennonite congress called in the Fall of 1918 looked for ways to deal with greatly-changed conditions in the Mennonite communities of what was now Soviet Russia. The congress hoped to take care of loose ends in the forestry and medical alternative service programs.

There was already some talk of emigration, and as Nestor Makhno's attacks began to increase, the larger issues of radical adaptations and survival needed the attention of the total Mennonite community for the next decade and more. There had been signals from the new regime that issues of conscience, such as military exemption, would not be ignored. For twenty years at least, options for an alternative to military service would remain, even under the new regime. But that is another story and must be told elsewhere.

[1] The phrase "for all time" appears in Point 8 of the Charter of Privileges obtained by the delegates Jakob Hoeppner and Johann Bartsch who, in 1786-87, negotiated terms of settlement for the Mennonites moving to New Russia. See the German version of the Charter in D. H. Epp, *Die Chortitzer Mennoniten,* first published in 1889 in Odessa, south Russia, and then republished at Steinbach, Manitoba, Canada, in 1984. The quotation is from page 21 of the republished version.

[2] The most complete description of forestry and medical service for the Mennonites in Russia is in Lawrence Klippenstein, "Mennonite Pacifism and State Service in Russia: A Case Study of State Church Relations, 1789-1936," (Ph.D. dissertation, University of Minnesota, 1984), See here p. 85. Unless otherwise stated, the material in this essay is drawn from that source.

[3] On Todleben's presentations to the south Russian Mennonites, see Klippenstein, "Mennonite Pacifism," 63ff.

[4] This was reported in a US Mennonite newspaper, *Zur Heimath,* (Sept. 21 and Oct. 7, 1881), 142 and 158 respectively.

[5] "Gross Anadolsche Forstei," *Die Rundschau,* (June 1, 1881), 2.

[6] A. Kroeker, "Wie sient es in der Welt aus?" *Christlicher Familienkalender,* (1904), 105.

[7] "Fortsetzung der Geschichte des Mennonitenvoelkleins in Russland waehrend des Jahres 1904," *Mennonitisches Jahrbuch,* (1904), 14-16.

[8] "Verschiedenes aus mennonitischen Kreisen," *Mennonitische Rundschau und Herold der Wahrheit,* (Nov. 2, 1904), 9.

[9] "Verschiedenes aus mennonitischen Kreisen," *Mennonitische Rundschau und Herold der Wahrheit,* (Apr. 8, 1905), 9.

[10] Jacob Hoemsen, "Als Sanitaeter im russisch-japanischen Kreig," *Der Bote,* (Feb. 8, 1972), 11.

[11] *Ibid.*

[12] The complete German translation of the Russian letter is in Heinrich Ediger, ed. *Beschluesse der von den geistlichen und anderen Vertretern der Mennonitengemeinden Russland abgehaltenen Konferenzen fuer die Jahre 1879 bis 1913* (Berdiansk: H. A. Ediger Verlag, 1914).

[13] Abram Kroeker, "Schilderungen aus dem russisch-japanischen Krieg," *Christliches Jahrbuch für Belehrung und Unterhaltung,* (1905), 121.

[14] Peter M. Friesen, *Die Alt-Evangelische Mennonitische Brüederschaft in Russland (1789-1910) im Rahmen der mennonitischen Gesamtgeschichte* (Halbstadt, Taurien: Verlagsgesellschaft Raduga, 1911), 521.

[15] Jacob Toews, "Zum mennonitischen Forsteidienst," *Botschafter,* (June 9, 1909), 1-4. Also "An den Herrn Neo-Mennonit," *Botschafter,* (Sept., 1909), 3.

[16] A. Kroeker, "Europaeischer Krieg," *Friedensstimme,* (July 19, 1914), 4.

[17] D. H. Epp, "In ernster Stunde," *Botschafter,* (July 27, 1914), 1.

[18] John P. Dyck, ed. *Troubles and Triumphs 1914-1924* (Springstein, MB: by the editor, 1981), Sept. 19 entry, 1914.

[19] *Ibid.*

[20] "Stadt und Landchronik," *Botschafter,* (after Aug. 26, 1914), 3.

[21] Heinrich Unruh, "Mennonitische Reservisten und Ratniki," *Friedensstimme,* (Aug. 13, 1914), 9.

[22] Johann Klassen, "Oberlicht," *Botschafter,* (Aug. 16, 1914), 1-2.

[23] D. Claassen, "Ein Abschiedswort an die jungen Männer in den Forsteien und an die Reservisten," *Friedensstimme,* (Aug. 16, 1914), 6.

[24] John P. Dyck, *Troubles and Triumphs*, entries for Aug. 23 and Sept. 1, 1914.

[25] P. A. Warkentin, "Erinnerungen an meine Dienstzeit als Sanitäter in den Jahren 1914-1918 im ersten Weltkrieg," *Onsi Tjedils: Ersatzdienst der Mennoniten in Russland unter den Romanows*, ed. by Waldemar Guenther, David P. Heidebrecht and Gerhard J. Peters (Yarrow, B.C.: by the editors, 1966), 86-100.

[26] "Aus deutschen Ansiedlungen," *Friedensstimme*, (Oct. 11, 1914), 6.

[27] H. Janz, "Rakitjaner Forstei den 10 Oktober, 1914," *Botschafter*, (Oct. 21, 1914), 4.

[28] J. J. Thiessen memoirs, typed manuscript in the author's files.

Photo Credits

1. Walter Quiring *Als Ihre Zeit Erfüllt War* (Saskatoon, SK: Modern Press, 1963), 177.
2. *Ibid.*, 169.
3. *Ibid.*, 171.
4. Gerhard Lohrenz *Heritage Remembered* (Winnipeg, MN: CMBC Publications, 1974), 190.
5. *Ibid.*, 174.
6. Walter Quiring *Als Ihre Zeit Erfüllt War* (Saskatoon, SK: Modern Press, 1963), 171.
7. Gerhard Lohrenz *Heritage Remembered* (Winnipeg, MN: CMBC Publications, 1974), 180.
8. *Ibid.*, 187.
9. *Ibid.*, 190.
10. *Ibid.*, 182.
11. *Ibid.*, 185.

Mennonite Alternative Service in Russia

Some personal experiences, 1911-1917

By Jacob Dick

Background to my Father's Story

It seems there is a weakness in the nature of the human race with respect to inter-generational interest. When our fathers wanted to relate their experiences to us, we were not interested. By the time we became interested, it was too late because our fathers had passed away.

The medical service travels and experiences related here belong to John Mathies (1890-1970), Nikolai Peters (1890-1954), Jacob Tiessen (1890-1969), and Abram Dück (1890-1970). John Mathies, Jacob Tiessen and Abram Dück grew up in the Schoenfeld municipality, east of the *Chortitza* (Old Colony) settlement. Nikolai Peters grew up in the village of Grigorjevka, southeast of Kiev.

These gentlemen (known in Low German as *tjiedels*) spent three years, beginning in 1911, at the Azov forestry camp near Mariupol, and for a considerable time served together on the medical services trains. My father, Abram Dück, wrote very little, but collected and cherished some one hundred photographs. However, his very good colleague John Mathies wrote quite a lot about their experiences. It is with John Mathies' writing that a common thread is woven into this document. Though the two men did not live through exactly the same events, their experiences were very similar, and their faces appear together on many photographs. John Mathies' diary and memoirs provide the first person narrative in the text that follows, in italic print.

My father-in-law, Peter J. Sawatzky, served in a phylloxera unit in the Crimea, near Yalta. His pictures and experiences have been added to this collection, as have some photographs from the collection of Peter Warkentin, my sister's father-in-law, who served at the Anadol forestry camp from 1905 to 1908.

Introducing Father and His Friends

Abram Dück

Nikolai Peters

Abram Dück was born in Schoenfeld, Ukraine, in 1890. He experienced a happy youth in that community. He enjoyed school very much and hoped to continue his studies in high school (*Zentralschule*). This was not possible because his older brother was married. His father's work as a minister meant that Abram had to look after the farm operations (*Wirtschaft*).

John Mathies was born in Schoenbrunn, Ukraine in 1890. Schoenbrunn was located only a few kilometres from Schoenfeld,

Abram Dück

Abram Mathies and John Mathies

Jacob Tiessen

which had the municipal office and the church. There John Mathies went to school and enjoyed a happy youth.

Nikolai Peters was also born in 1890 and spent his childhood years in Grigorievka. During his teen years he helped his widowed mother operate the family farm.

Jacob Tiessen was born in Schoenfeld in the same year as the others, and grew up and went to school there. John Mathies married Agnes Friesen, also from Schoenbrunn, on September 1,1918. Abram Dück married Maria Driedger on August 25, 1919 in Schoenfeld. Nikolai Peters married Sarah Klassen from Nikolaipol in 1921. Jacob Tiessen married Anna Janzen of Blumenfeld on July 26, 1931.

"Sending you a picture of two soldiers. You will recognize one; the other is a Peters. You can see how we look. It's not the greatest, but that's how it is. We are having a birthday today and will have something to eat. The birthday child has brought the food."

Nikolai Peters and Abram Dück

ForestryService

The *Forstei* (Forestry Camp) at Azov

Photos above and below:
Buildings at the Azov *Forstei*

Photo left:
Head Forester V. Sijakov and his daughter

De Opnom (The Reception)

In the fall of 1911, John Mathies and Abram Dück had to report to the civil authorities in Gulaipole of the Jekaterinoslav province. They were found to be healthy and ordered to serve their time at the *Forstei* Azov, somewhat north of Mariupol on the Sea of Azov.

On March 1,1912 these colleagues left Schoenfeld to travel to Gaitschurr, then to Pologi and finally to Rossovka. From the train station it was still another thirty kilometres by horse and wagon to the Azov forestry camp. About sixty new recruits had already arrived at this *Forstei*. Having safely found their destination, they were greeted and immediately separated from each other.

Initiation to camp life.

"Sincere greetings, my dear friend from Vladimir. You get a small idea of a welcoming 'initiation.' Give my greetings to Miss Sara and all the other ladies. From your, Herman Enns."
November 12, 1908

The new recruits were placed in several wings of the camp barracks. Each wing housed thirteen persons, and six new recruits were placed with seven older "colleagues." The recruits were given the necessities for making their beds—a bag of straw and a rope to attach to the bed frame.

Then the "initiation" of the new recruits continued. The filling of the sleeping bags with straw was meticulously supervised by the more senior unit members. Some new recruits were ordered to put more straw in the bags, while others were ordered to remove a little. Only a very precise amount of straw was acceptable. Some recruits were reprimanded that their needle for sewing up the bag was getting much too hot and so had to be ducked in a pail of water for cooling.

The new recruits were also required to serve their colleagues coffee in the morning. They had to obtain hot water from the kitchen before their seniors were required to get up. John Mathies was destined to stay in the first wing of the dormitory. It was called the "Chicken Coop." Abram Dück was welcomed into the "Monkey Cage."

Tree Planting and Forest Management

Work in the forest was located approximately four kilometres from the main camp. It took the men only twenty minutes to reach the camp when returning for the noon meal, for which they worked up great

The South Coast camp near Yalta, Crimea, August 14, 1911

Marching to work

We will get through these years somehow. I am sending you this card so that you will have some idea how we get to work. I am near the front. Only my head is visible because I am on the far side. In closing, I greet you once more with kisses. From your loving and "not-to-be-forgotten" son, Abram.

Tree planting

Abram Dück, John Mathies, and Nikolai Peters are part of this group.

A Sunday afternoon

"There will be no more apples. Hopefully by then we will have peace. It is high time or we will forget one another. In closing I greet you fondly. From your loving brother Johann. Greet J. Friesen and any other acquaintances."

Photo left:
Middle row, second from the left: Abram Dück.
Middle row, third from the left: Nikolai Peters.

"Sending you a group of gentlemen which you probably all know. Greetings from Abram Dück. Until we meet again."

Middle: John Mathies
Far right: Abram Dück

appetites. Their senior supervisors (the *Gefreite*) were J. Warkentin and G. Schroeder.

A tree nursery existed at the Azov *Forstei*, about half a kilometre from the camp. John Mathies felt lucky to have been assigned to the tree nursery and wrote, "There is much I could learn working here." Nikolai Spenst from Memrik was the *Gefreiter* at the tree nursery. "He was a good guitar player and had a love for life. He was a blessing to many of us."

In the summer of 1914, business manager (*Oekonom*), Bernard Fast, left the camp for his home in Terek. Mr. G. Klassen from Davlekanovo then became the new *Oekonom*. Mr. Fast was a much-loved chaplain. John Mathies had the privilege of taking the Fast family to Mariupol where they caught the train. "They were loved and cherished by us all," writes John Mathies. Tragically, Mr. Fast and some others were later killed by a band of robbers.

Working in the forest

Photo above:
Working in the forest

Photos right and below:
Harvesting trees

Threshing machine at Azov
John Mathies on left

Funeral at Azov

Johann Goertzen
"On March 4, 1913 our colleague was buried. He was in his second year of service."

In his third and last year John Mathies was privileged to work with *Oekonom* Bernard Fast, who had also served as camp administrator. John Mathies frequently went with Mr. Fast to Mariupol to purchase goods for the camp.

Music was a very important form of entertainment. Instruments included guitars, accordions, accordion with two rows of keys (called *tvee reye* in Low German) and a full band complement.

Third from left: Abram Dück
Third from right: Nikolai Peters

May First Festival

A band played at the First of May celebrations and other special occasions. Abram Dück played in the band all three years he spent at the *Forstei*. He enjoyed music very much. Nikolai Peters also played in the band.

Interlude for a photo

While serving on the medical trains, Abram Dück's train came to a stop in a city in the Caucasus on Easter Sunday. The *Sanitäter* (medics) decided to visit a local Lutheran Church to mark this important Christian holiday. Abram Dück fondly remembered the service which included a ladies' choir. He wrote that the singing was so wonderful, and the ladies' faces so beautiful that, "I thought I was in Heaven."

P. Albrecht

In some forestry camps the music developed to a very skilled level. The head Forester would hire a band conductor to instruct the young men to play various instruments. They also got time off work to practice their instruments. This resulted in local concerts at the *Forstei,* as well as programs in some of the villages near the camps.

Other Entertainment

The total scope of entertainment forms is difficult to determine. Entertainment was more loosely defined at the camps, to include many forms of relaxation, as photographs from the time reveal.

Photo above:
An automobile adds interest to this group photo

Photo below:
A large picnic in the woods

Abram Dück

Visitors at the camp

Relaxation on a Sunday

A picnic with visitors

"Medics on the Medical Train No. 208 ARZU at the ruins of a fortress near the Station Ksanka. View of the insides of the ruins."

Photo right:
How did the ladies get up there?

Exercising

Even though the men worked hard physically, there was always time for fun. Later, some of the conscripts who had received more formal education instructed their colleagues in calisthenics.

Playful in the snow

Gymnastics at the Forestry

Peter J. Warkentin on the right

Road Building

Roads were built by special units. This was an activity for which the young men could volunteer. A number of these recruits worked in the Crimea.

Road building

Photos above and right:
The South Coast forestry camp, Crimea
Photo above:
Henry Warkentin and Peter Sawatzky

Photo right:
Peter Sawatzky

A phylloxera unit was organized to help control insects in the vineyards of the southern Crimea. The units lived in tents, which were rather primitive, and the work required moving around to the different areas. It consisted of searching for a very small insect called phylloxera which had infested the vineyards. To try to prevent this insect from establishing itself on the plants, the roots of the plants were dissected and examined under a magnifying glass.

Picnic with the Forester's auto. Peter Sawatzky is on the far right.

Working near Yalta, Crimea. Peter Sawatzky is on the right.

Since the men had no training for this work, they were sent to Odessa where infestations of phylloxera were also found. This provided an opportunity to become familiar with the insect. Smaller groups were sent out to the owners of the vineyards to instruct them in the spraying of the plants. The volunteers got to see the wonderful scenery of the Crimea, enjoyed the mild weather, and saw the magnificent palace in the city of Lividia, the glorious horse ranch, and the luscious fruit gardens in the vicinity. "The years at the *Forstei* were the best in my life," wrote John Mathies as he looked back on those days.

Pheasant farm building used for shelter

"This group served in the South Coast forestry camp approximately three miles above Yalta on the Black Sea. This unit consisted of reservists who were recruited in 1915 during the war. Housing was in barns that had been constructed for breeding pheasants."

Medical Service (*Sanitätsdienst*)

When the war against Germany and Turkey broke out in 1914, fifty men immediately volunteered with the Red Cross to become *Sanitäter* (medical service workers). Abram Dück and John Mathies were part of this group. The Russian government was very happy about this gesture and ordered the volunteers to report in Jekaterinoslav on October 17,1914.

When Abram Dück informed his parents, Rev. and Mrs. Jakob Leonard Dück, about his decision to volunteer for the medical service, his father expressed his concern about the decision. Still he wished him God's blessing and promised their prayers would go with him.

On October 16, 1914, the volunteers left camp. Horses and wagons were obtained from a nearby Russian village. The head forester V. Sijakov, in his uniform, and Sergei Mukotivich, his assistant, had come to see the group depart. Sijakov placed the volunteers before him in a straight row while the others were placed on either side. He praised the group heartily, after which the band played "God protect the Tsar." As the wagons started leaving the camp, the somewhat emotional representatives and others in the crowd raised their hats in admiration. Some tears were also shed on this occasion.

Johann Wiebe

"To: Abram Jacob Dück
To remember your friend
Johann Johann Wiebe."
February 22, 1912

The three hundred or so men from the different forestry camps who had arrived in Jekaterinoslav were greeted in a friendly manner. However, they were also informed that until now they had been under civilian authority, but from now on they would be under military authority. Such a regime, they were told, would be much more regulated and rigid.

Our Red Cross uniform.

John Mathies goes on to tell us more about their experiences: *Now we were on the way to Moscow where we were placed in the All Russian Zemstvo Union (ARZU, the organization which set up and administered the work of Mennonite medics in WWI). The men from the Azov* Forstei *were assigned to Trains No. 205 and 206. For twelve hundred wounded soldiers there were two doctors, six nurses, and one superintendent over us medical personnel. We also had one administrator, one secretary, three cooks and forty other medical persons as part of our staff.*

In Moscow we also received our military uniforms, which clothed us from head to toe. We displayed a red cross on our caps and on one arm of our leather jackets.

Interior Service

We worked in these Red Cross trains for only a short time. Our task was to move the wounded from hospitals in Moscow to Vologda, Vjatka, and Perm, over the Ural Mountains to Nizhni-Tagil, nearly two thousand versts *(two thousand kilometres) to the northeast. On arrival, we were greeted with band music. The largest iron mountain in the world is purportedly located here in Nizhni-Tagil (See Map 1).*

On the return trip, all the train cars were disinfected with steam and formalin to kill all the germs, including lice. All the twelve beds were re-cleaned and all the dishes washed, because we never knew when an inspection might take place.

On one return trip, we observed God's handiwork when travelling through the Ural Mountains and looking at the wonderful forests. At this point, the loveliness of God's handiwork seemed so near. When wandering through the train, we could see some of the men writing letters, others studying their Bibles, and still others fetching their tea or coffee from the kitchen. This created closeness among us fellows, as well as much laughter.

At the end of the day, we said our little prayers, went to bed and fell asleep to the "tick-tack" of the wheels of the train. It was a long trip back to Moscow. We made several more trips from Moscow to the Volga, a distance of about seven hundred versts (seven hundred kilometres). We also made some shorter trips. Then Trains No. 205 and 206 were turned over to the Polish volunteers.

Left to right: Jacob Tiessen, Abram Driedger, Abram Dück, and David Dück.

Medical Train No. 211

A station stop for the train.

Three new trains, Nos. 207, 208, and 209, were prepared at this point. John Mathies was ordered to Train No. 209, and Abram Dück to Train No. 208. These three trains were prepared with all required supplies, doctors, nurses and Red Cross personnel, less the cars for the wounded, and sent to the Turkish front. All personnel were vaccinated for the various diseases prevalent in the Caucasus area, especially malaria.

On December 21, 1914 we left Moscow to go to Rostov-on-Don, where the weather was already much nicer. From there we went to Mineralny Vody, Grozny (now in Chechnya), Khasavjurt on the Terek, Petrovsk, and Derbent, a city surrounded by a wall, on the Caspian Sea. Beautiful gardens extended to the seashore, and the air was springlike. We could still see the Caucasus mountains, Mount Elbrus (18,482 feet) which we had passed some time ago, and Mount Kasbek (12,886 feet). Oh, how wonderful God's creation is! We travelled further to Jadsharg and to Baku. Baku is a great industrial city, with many oil wells where naphtha, gasoline, and other petroleum products were made. It lies on the Caspian Sea.

View of the Kremlin from Moskvoretz Bridge in Moscow.

General view of the Kremlin

Further on lay Aluatu and Elisavetpol with beautiful vineyards. Many German Lutheran and Evangelicals lived in this area. Then we went on to Naftlug and finally to beautiful Tiflis. It was December 30, 1914 and we had travelled 2900 verst *(2900 kilometres) altogether (see Maps 2 & 3).*

In Tiflis we were required to fit up the railroad cars for the transportation of the wounded. With Train No. 209 ready, we left for the Turkish front. We drove to Naftlug and Sinain between the ever-higher mountains. Our train had two locomotives in front and one double locomotive on the end. It was really awesome to hear the echo of the train whistles in the mountains. We already felt the change in the weather, and snow was becoming visible on the mountains.

We travelled through several small tunnels and one very long tunnel where everything became dark. After passing through the tunnels, we came to a snow-covered valley and started wearing the wool-covered caps that we had received in Tiflis. The mountain had a height of 8989 feet. We were nearing Alexandropol, an important transportation centre. From there we travelled southeast toward the Persian border, to

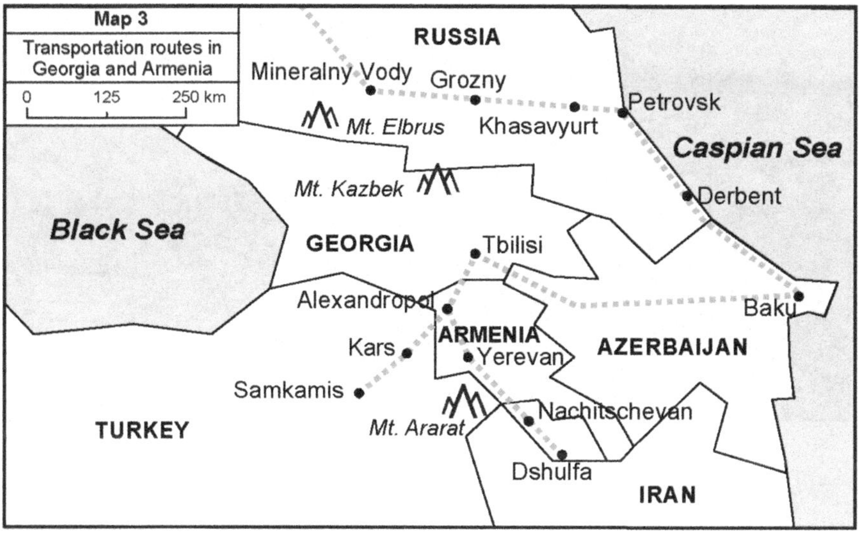

Etschmiardsin, Uluchallu, Yerevan, Nachitschevan and Dshulfa, on the border of Persia at the Arak River. The Arak River flows past Mount Ararat and forms the Russian-Persian border, then on to the Caspian Sea (see Map 3).

We travelled further to Kars, a small fortress, then Begliesschnett, where the Mohammedans live in earthen caves, and finally arrived at Sarakamis, the location of the Russian-Turkish front. There we were at an elevation of ten thousand feet and it was bitterly cold.

At that location we saw many corpses all around. Two Turkish corpses had been laid at the end of the tracks close to the station, frozen into the ice and could not be buried until spring. This death and sadness causes one to think seriously about the reasons for going to war. In wartime, parents, mothers, and wives wait, and none know what has happened to their loved ones.

"Dear Family and Relatives,
I am sending you a postcard of our train taken by myself on the way from Tiflis to Alexandropol in the high mountains near a smelting factory. It is a beautiful area. I am, thank God, alive and well, which I wish you as well. There is nothing new to report. Best wishes to you, your loving son. Goodbye. I have not received a letter for a long time."

Bridge over the Arak River

We then collected many wounded, mostly of Turkish descent. Many had frozen hands and feet, and the smell from pus was terrible. Many died on the way to Tiflis three hundred versts *(300 kilometres) away. We*

Medics getting well deserved fresh air

Disinfecting the trains

"Disinfecting Medical Train No. 208 at Baku
This is a picture of the disinfection operation. Alongside the train you see a large pipe, the main pipe, and smaller pipes going to each car and entering through the windows. The main pipe is connected to a generator from which it is directed into the cars. Eight cars are disinfected at a time. It takes about twenty minutes. The temperature in the cars is about sixty degrees centigrade. All the insects perish."

cared mercifully for the wounded and unloaded them at the hospital, where many had their legs amputated.

In Naftlug we loaded more wounded soldiers and travelled to Baku, Baldshary, Khasavjurt on the Terek, Prochladnaia (where some Mennonites lived) Suvarovka, Armavir and Rostov-on-Don. This trip took over seven days and nights. It was not always easy, but I was young and healthy, and happy that I could do this. I carried many a wounded or diseased person on my back. They were all so happy, whether friend or foe. To me they were all friends.

The wounded were transported directly from the battle area on stretchers made of two bamboo poles. The poles were fastened to the saddles of two animals, either horse, mule, donkey, or buffalo. This was necessary because the wounded were transported on extremely narrow paths through the mountains where the animals had to walk one behind the other.

"Sending you a glimpse of our work. I photographed this as we were loading the sick. The sun was shining directly from the side and therefore the faces are not clear. The cross indicates the stretcher with the patient lying inside. This stretcher is carried by two horses. These are only for the seriously injured. The less seriously injured are brought on wagons. The double crosses indicate a covered stretcher. You can see the rear horse. Its head is on the other side of the stretcher. In the background are the mountains. Only the top of the train is visible. Another time I want to photograph the wagons."

I am so thankful to God that I had the opportunity to do this work. I like to believe the Mennonites were called to do this task and that they carried out this task as medical personnel in a trustworthy manner.

A multi-track railroad

View of Mount Ararat from Yerevan

On one occasion we saw a train full of wounded Turkish soldiers standing on another track near the hospital. There were no Mennonite workers on this train, and at each door was a soldier fitted with weapons.

Unmindful of anything, I looked into one of the cars. Corpses were lying around in the wagon on straw. One wounded soldier begged for water, but the guard totally ignored him. Then a scantily clad wounded soldier jumped from the train car and knelt at a puddle of fresh rainwater to drink. He was immediately butted with a rifle and ordered back into the car. I had no weapon, but we had a heated discussion in which I demanded he give the wounded water for they were thirsty.

We travelled the stretch from Alexandropol to Dshulfa often. It was always a long, barren, desert stretch with tropical heat. We picked up more wounded and sick who had already been brought as far as the Urmia Sea and the Van Sea. Our Mennonite boys faithfully served, with autos and trucks, in bringing the wounded this far. This area was somewhat dangerous because of the violent nature of the local inhabitants, the Mohammedans. Not all of our colleagues returned safely. A Mr. Goossen, also from the Azov Forstei, *lost his life here.*

When we were in the area of Yerevan and Nachietschevan and passed by Mount Ararat, said to be 16,946 feet high, we often thought of Noah. Had Noah not lived in this area with his sons? We also visited Noah's

Noah's Grave

Astafiev Street in Yerevan

grave, some thirty versts *(thirty kilometres) from the mountain. It was located in an Armenian cemetery surrounded by a stone wall. There were two gates in the wall with an altar in the middle, indicative that pilgrims still went there to worship and pray.*

On Easter Day, 1915, our Train, No. 209, waited in Yerevan for orders. The sun had just risen and Mount Ararat was clearly visible. We could

Station stop

even see the snow cap. We probably were some sixty versts *(sixty kilometres) away but it seemed so close. Some of our colleagues were outside enjoying the fresh morning air, and those inside the train looked out of the windows.*

Mr. David Sawatzky (currently living in Vineland, Ontario), was outside the train that morning. He and his companions were surprised to hear the Easter Greeting, Christos Voskress *(Christ is arisen), and to receive three kisses with the greeting* Vo Istinu Voskress *(He is truly arisen). This also happened to others who were standing in the row. The Governor of Yerevan and his generals were there to see our medical train, No. 209, and were the source of the greeting. The governor's Easter greeting was repeated many times as the group made its way through the train. This visit from the Governor and his personnel cheered our spirits and we appreciated it very much.*

The local headquarters office of the ARZU (All Russian Zemstvo Union, the organization which set up and administered the work of

A little visit during a train stop

Dressed for high mountain weather

"In this picture we are getting pancakes."

Mennonite medics in WWI) was in Tiflis, the source of our orders and supplies. We were at the Persian border in Dshulfa, which had a big, beautiful train station. There were few trees in sight, except for a few big ones at the station. It was very hot. To the south, in the distance over the Arak River, which formed the border, were high mountains. Near Dshulfa to the north, on the Russian side, were beautiful hills. One could see herds of sheep and goats, that probably grazed the mountains.

Two Red Cross men were to go to Tiflis, with the mail train, to obtain leather jackets and boots for our colleagues. Mr. D. Boldt and myself were elected. As we waited at the tracks for the train, we noticed a young couple. A young man, a train employee in uniform, was with his wife, who was totally veiled except for her eyes. The couple were accompanied by a man who was probably the woman's brother. The father gave them the mullah's blessing with the "Allah–Allah" before their departure. We boarded the train along with this young couple.

We had the good fortune to observe this happy couple all the way to Tiflis. As the train started to move, the veil fell from the young woman's head. I then thought of Moses who, in the wilderness, met the high priest's daughter, Zippora, whom he married. When we got off the train in Tiflis, we wished the young couple much happiness. I hoped that I too may find such happiness. The Tiflis area had many German Lutherans and German Evangelicals in the village where D. Boldt met his sweetheart. I later visited this couple in Neukirch, Molotschna, and believe they are also very happy. When we returned from Tiflis to our Train, No. 209, we were happy and fortunate to rest with our colleagues again.

We decided to go swimming in the cold, foamy Arak River. We were warned not to go very far into the river, as horses and soldiers had already drowned. This did not seem dangerous to me, so I went in a little deeper, not realizing that I was already being gripped by the current. I had not grown up on the Dnieper River, so I was not aware of the power of river currents.

Photos left and below:
A view of the ruins near Ksanka.

Medics on Medical Train No. 208
At the ruins of a fortress near Ksanka. View of the inside of the ruins

Before long, my colleagues were out of sight and the river became wider and wider. I feared I might yet reach the Caspian Sea. I tried to rest on my back, but this position was not comfortable as I had already choked several times. I noticed a small something on the Russian side. I tried to swim again, this time slightly into the current, and managed to save myself. My chest and sides were bloodied from the rocks along the shore, and I had yet to climb the thirty-foot height of the rocky shore. The stones were hot like fire. I thanked God for saving my life. I had already been swept one verst *(one kilometre) downstream, and no one thought that I was still alive.*

The area west of Tiflis, Mzchett, Ksanka Gorgi and Batum, is also beautiful. The area around Borshomi is similar to Switzerland. The wealthy estate owner, Nikolai Nikolaovitsh, owned seventy-five thousand dessiatins *(202.5 acres, or 81.9 hectares) of land in the area. Much nobility from northern Russia lived here in the Caucasus. There were many mixed races included Tatars, Armenians, Georgians, Ossetians, Kurds and Turks. The people were dark in complexion and small in stature. I was always astonished when meeting them on the street by how quietly they walked in their light leather boots, as if in stockings.*

The Arak River

Cakvo Tea Alley in Batum Centre

Mario Plaza in Batum

Green Peninsula in the Batum area

Green promenade in the Batum area

Boulevard in Batum

Many of the *Sanitäter* became infected by the diseases that were rampant at the time. Abram Dück became infected with malaria, was hospitalized, then released for recovery, and spent some time at home. He was very weak when released from the hospital for recovery at home. He said that if it had not been for the good will of fellow passengers, he would have succumbed somewhere on the way.

John Mathies' account continues: *Malaria was prevalent in the area, and many of the medical personnel became ill. Our dear colleague, J. N. Goertzen, was taken to the Alexandropol hospital and died there. J. P. Thiessen became ill near Sarikamis and died there.*

In November, 1915, our train, No. 209, stood at Ksanka at the Kura River. It is a tributary to the Arak River, and is a magnificent area. It is there that Noah and his sons are said to have lived and built the city of Mschett.

I became ill with typhus and malaria, and was carried by my colleagues in a Bahre *(stretcher) to a car for the sick that was added to the mail train. I was taken to a hospital in Tiflis where I was treated for some time: I was able to withstand the typhus but not the malaria. Our Jewish doctor, Mark Naumovich, was also treated here. He was located in the next wing from me and I would sometimes visit him in my stretcher. He would generally express his feelings with "Oh God, why is it so difficult?" This kind Jewish man later worked on Train 202 for a time and gave me extra furlough, because I had given my furlough to my brother, David Mathies, who was already married and who needed it more than I did.*

I required a different climate, and was thus allowed to travel back to Moscow. I was given an assistant, since I could hardly walk. In Baku we had to change trains and went into the train station, where we were quickly joined by two ladies who were also travelling to Rostov-on-Don. They immediately recognized that I had malaria and wanted to accompany us. I thanked them for their thoughtfulness and in no time they had obtained a bitter root from the apothecary near the station which they called Goritchaka. *They cooked the root and gave me a big bottle to drink. In Rostov, where we parted, I had recovered to the point that I no longer needed my assistant.*

Baku railway station

Triumph Place with Bouff Theatre in Moscow

When we arrived in Moscow we each took our own taxi. It had changed from summer into winter, and was already snowing. Sredni Tishinskii Pereulok (Tishinski Street) was a welcome spot for us. Mr. Eitzen greeted me and on the following day brought me to the Arbat hospital, which was familiar to many Sanitäter.

Forest watchmen

There I also met Mr. Schroeder (at the time of writing living in British Columbia). He had also arrived from the Caucasus with Train No. 208, where he had been an administrator. He had also been a forest watchman at the Azov Forstei.

He had his own horse and a cook, named D. Balzer. They both lived by themselves and had a good deal of freedom. In the Arbat hospital in 1915 I contracted chicken pox, so I was sent with others fifteen versts *(fifteen kilometres) to Sokolnikii, a state hospital. Only patients with chicken pox were admitted to this hospital, and many died. I immediately came to lie beside a Mr. Derksen from the Kuban (who, at the time of writing, lives in Leamington, Ontario). I stayed at this hospital another month and then received a month furlough.*

In the hospital in Tiflis my life was in God's hands. When I became conscious, a female doctor and two nurses rubbed me with vinegar and continuously called me and advised me not to fall asleep, as that would be my end. That is where I realized that people with darker skin are also respectable people with sympathetic hearts.

In Moscow, in the state hospital where only small pox victims were treated, we were also treated very well but still many died. The hospital had two big wings with a large open door for men, women, and children.

Some patients had only one big pox while others were full of pox. In the morning men would gather the corpses and carry them to Trog *(the morgue). Through all this the people remained happy, and sat together, played cards and waited for their freedom. We were not to wash ourselves, because washing resulted in scars. I wrapped my hands with a towel for the night so that I could not scratch my face, and was thus spared scars. Some could not keep from scratching themselves and so were scarred.*

In 1916 I was lucky enough to be assigned to Train No. 202 in which my brother David Mathies also served. Other Schoenfelders on the train included Abram Dück, Henry Enns, Peter Fast, David Baerg, Henry Riesen, Peter Thiessen, Isaac Sawatzky, and later Kornelius Neufeld. In Moscow, we loaded and transported the wounded in northerly and southeasterly directions over the Volga and further. There was considerable traffic on the Volga. Most cities and towns and their markets looked similar to Baku, Kasanj, and Jaroslavl. Once while we were stationed at Saratov, where many Germans lived, we also travelled on the Volga. It was a wonderful trip.

At one point we were stopped in Samara, where we had unloaded the sick and had some spare time. My cousin, David Baerg, and I were walking in the city park on the tracks of an electric street car, when we

Lubiansky Place, Moscow

became aware of the fact that the car was approaching us from behind. We jumped out of the way, but the cow catcher hit the heel of David Baerg's boot and threw him thirty feet, without any serious injuries.

On another occasion in the Caucasus region, when it was very hot and muggy, we found a comfortable place on the roof of the train, near the chimney. We had to be careful when passing under bridges and under telegraph lines, and had to duck quickly to get under them. Danger lurked nearby. My colleague Krause was hit on the head and with a cracked skull fell between the cars and through the bridge supports into a small stream some thirty feet below. The Russians rescued Krause and took him to a hospital. After several months of recovery, Krause rejoined us.

In 1916 many of the older generals were replaced by younger ones. The older generals were placed in areas that were less strenuous, like hospitals, trains, or even on the medical trains. We also received one such general in Train No. 202. He weighed three hundred and fifty pounds and on his first trip he already had our boys arrested. Guards were placed at his door, our hair was to be cut, and all rules were to be memorized. The train engineer was to be arrested because the general's tea pot had fallen from the table.

The entire train personnel were enraged at our misfortune. All personnel, including our doctor, Lady Eugenia Stepanovna, wanted to leave the train. It did not take long until all those arrested were released. Our hair was no longer required to be cut in a crew cut, and the old general became our best friend and sat with us. He apologized very much and said he had not been aware that he was dealing with intelligent people. Because of him we received more furlough time.

In 1917 things changed again. The old generals were released, and our Pluganov was also let go. Upon leaving, he was very emotional, and those who carried his luggage learned his feelings. We in the Sanitätsdienst *had learned to respect him; of this he was well aware and deserving.*

In our younger years we are sometimes prone to act without very much thought. I had already served on Train No. 202 for over a year when I had the opportunity to join my Asov colleagues in the Caucasus again and became a Sanitäter *on Train No. 209. Dr. Paulina Nikolajevna Vogel was again glad to have me on the train. In my absence three more Schoenfelder had joined the group: Mr. Abram Rogalsky, (at the time of writing, of Winnipeg, Manitoba); Mr. Gerhard Schroeder, (at the time of writing, of Leamington, Ontario); and Mr. Jacob Enns, (who has since died in Russia).*

Because of the revolution, everything changed. The All Russian Zemstvo Union was rich and very active during the war. Our Sanitäter *also became aware of easier work in hospitals, orphanages, bakeries, and even ship building.*

I became involved in shipbuilding and was temporarily stationed in Tiflis with two other Schoenfelder, John Friesen (now of Winnipeg) and Jacob Tiessen. From Tiflis we went to Batum, then on a big war ship with over a thousand military troops. Without lights and during the night we left for Trapezund, Turkey, escorted by smaller war ships. Early in the morning we arrived in the harbour in Trapezund. Barely had we arrived when we were walking in a narrow street and bombs were falling. We had to scurry into basements for shelter together with the Turks. There in Trapezund we met other Mennonite Sanitäter*, and chauffeurs who brought the wounded to Sarikamis from Erserum and Olte.*

We went in a small motor boat some thirty versts *(thirty kilometres) to Surgonne to a small shipbuilding enterprise located on the shore of the sea. All around were high mountains with valleys and narrow roads. The Turks lived in nice houses, all built of walnut. The lumber was always sixteen feet long and at least one foot, and sometimes two feet, wide.*

War brings with it such destruction. It is better to tear down everything than to burn it, because the poor people can then still find something to build for some shelter. There was a Turkish barber, a friendly older gentleman, who still had a small business. Every day much military activity took place, but it was still peaceful. One day two soldiers entered the barber's shop, started looting, broke his mirror, and even hit him so hard he was bleeding. I thought to myself: why is no one saying anything and why must I put up with it? I ran to the barber's defence and pushed one of the soldiers down, only to see the other one coming at me. I wondered how long this could go on. I did not see these Turks again. When people talk of pacifism, I often wonder if most people actually know what real life is all about.

I worked on the small boats on the shore of the sea and was able to observe a good deal. Near the mountain, there was a mosque where Turks worshipped, and I visited it. Before the entrance were several water taps

Mosque in Yerevan

for washing, and a place for removing shoes. Worshippers crawled on hands and knees over a beautiful carpet, to His Holiness with a red cap on his head.

At that time, Turkish men could have several wives. It was interesting to see the wives carrying corn straw on their backs, with their heads and faces covered to the eyes with veils. The men followed their wives, wearing red-coloured beards, red-coloured finger nails, and red-coloured stockings. Russian-German military marched past singing, "If I should die in a foreign land, complain to me alone." The air was filled with filth, and the Turkish front was approaching. Bread was already being rationed, and the fish soup was becoming ever thinner. No one knew what would happen.

When I was working on a motor boat and it was becoming twilight, I noticed that a small boat had landed on the beach with five armed military men. The captain and boatmen were Ossetians, while the two sailors were Turks. All were Russian citizens. The mechanic was a Mennonite from the Kuban, named Alexander Bergmann. His assistant had become ill and they needed a replacement. The situation was as if planned! I acquainted myself with Alexander Bergmann and in no time packed my suitcase, put it on the boat, and loaded the ship with items destined for Batum. We sailed at night, and I was thankful to be able to leave that place. The ship had sails, and a motor was also attached.

Tiflisskaja Street in Batum

Tea-drying estate of N. Sabajev in the Batum area

Along the shore of the sea were places where we loaded and unloaded grain, military provisions, gasoline, and passengers. Batum had a big harbour, good transportation and streets lined with palm trees.

Our ship always leaked water into the engine room, which had to be pumped out from time to time. Debris forever collected under the engine and would float on the water like oil, as did gasoline, which made for a very dangerous situation. There was no electricity on the ship, and so one day a fire broke out because of our candles. It was our luck that we were still in the harbour and could extinguish it. On another occasion, while at high sea, we ran into a severe storm which no one thought we could survive. Our engine was old, the magneto system worked poorly, and so everything stopped. The sails had to be raised again, and since the ship could not be controlled without the propeller, we drifted at sea. When the storm finally let up, it took us three days to get back into the harbour in Batum. In my mind I can still see the snowcapped mountains of Turkey. It was December of 1917, just before Christmas, when I decided not to sail any more.

The people in the streets of Batum were restless. Soldiers were starting to sell their clothing at the market, and many were arrested. Our ship required repairs. I bid my dear friend Alexander farewell and left with the captain to a restaurant for a coffee. The captain told me a secret: he had no intention of returning to the ship either.

Through all these changes, I became ill and went to see a doctor who admitted me into the hospital. At this time, many men came before the commission and were discharged, as was the case with me. We had to

Boys' School at Batum

By the sea shore at Batum

present ourselves before the Minister of War and received a free ticket home, which took us by ship from Batum to Novorossiisk and by train the rest of the way home.

When we departed Batum in January of 1918, things were already very chaotic, and gunfire could be heard in the streets of Batum. From Novorossiisk, we went to Krasnodar and Zaritzyn (later Stalingrad, now Volgograd). We took a long, roundabout way to Lugansk, then went to Tschaplino and Gaitschur.

There was a front line at Rostov, where we were fired at. The trains were full of military personnel who had not surrendered their weapons, and this was the reason they were shooting at us. Not everyone returned to their homes, but in January of 1918 I arrived in our beautiful Schoenbrunn. We had barely arrived home when the bandits began looting our homes. Life was no longer safe. Was this our reward for serving our country faithfully for six years? But God had protected us from so much, it seemed as if that difficult time had all been a dream.

In the spring of 1917, the year of the Revolution, our Train, No. 202, was stopped north-east of Moscow in Jaroslav when Kerenski, with his staff in hand, stopped at the train station and declared freedom. Within minutes the station was full of his recruits. Similar scenes were taking place throughout the country. We received uninvited guests: labourers from the coal mines, and even criminals that had been released. We Mennonites, however, held together, and even moved closer together, since we were not safe alone on our individual farms. Through all the turmoil, people still got married.

After Abram Dück had fully recovered from malaria, the collapse of the Russian government was already under way. He never returned for further service, nor was he ever properly discharged.

Additional Photos

Summer uniforms of the forestry camps

A typical group pose

"To Abram Jacob Dück, Remembering our times together. From Johann Abr. Koop, Postal Station Svidovka, Village Kotlarovka, Ekaterin Gub."

The cook

To Abram Dück,
Wiesenfeld

"Dear Friends,
Since I have an opportunity to send some post cards cheaply I am taking advantage of the situation. I am well. I'm sending you a photo of our threshing operation. The workers and I are among the spectators. Greetings from your friend."

Herman Enns
Aug. 10, 1914

William Dück

Abram Driedger

"Dear Friend: Greetings from afar. Wishing you success in the threshing. Has it almost ended? Why is no one writing? Abram does not write at all. Please do write. In closing, greetings again."

Abram Driedger

Photo below:
A popular instrument was the accordion, also known as a "two rower."

Photo above:
Making music

Ruin near Ksanka

"Fortress in the Caucasus"
-Abram Dück

"To remember our times on Train No. 208."
J. D. Fast
Feb. 22, 1917

Johann Wiebe

Tavr. Gub. Post Office, Verosni Rogatchik

Kol Alexandrovka

"To our good times together. From Johann Heinrich Wiebe."

Johann Cornies

David Cornies

Johann Cornies

Peter Cornies

G. Loewen

Village Lesnoe
P.O. Snamenskaya
Tomsk Gub.

Nikolai Driedger Herman Enns

Poem July 16, 1915
"Souls in judgement
Strangers of the World
Have Sufficient of God's Grace
Soon it will be better
A comfort and reminder for us all
Your, Nikolai"

Left: Nikolai Driediger
Right: Abram Driediger

Herman Enns

Jakob Dück

“To remember me, your dear friend P. I. Cornies”
Tiflis, May 1, 1915

Jakob Dück

Both photos taken in the same studio

"To: Mr. Peter Johann Cornies, remember your friend Jakob Franz Wiens"

Jakob Wiens

Willi Penner

Aaron Tiessen

George Heidebrecht

Johann Friesen

David Dück   Abram Dück

"Here come guests. If you don't know these guests, then write who they are. A greeting and a kiss, from your son, Abram. Until we meet again."

Jakob Heidebrecht

Peter Heidebrecht

Jakob Driediger

"I am sending you this photograph of myself for remembrance. I would gladly send all my siblings a picture but that is too expensive. What are you doing? When 'Herring' must leave will you become the machinist? Did you make faces at my John while he was being photographed? He is turning his face to one side and has his eyes and mouth wide open. Teach him to run and when I come home he will come to meet me. Closing with a hearty greeting, your brother Jakob. P. S. Have just finished my noon day nap."

Photo Above:
Abram Warkentin

Photo Below:
In the forest

Samara Gub., P. O. Petanov, Village Bogomazov Abram P. Reimer

Abram Reimer

"To the good memory of your friend W. J. Boldt." P. O. Slavgorod, Tomsk Gub., Village Korotenka

G.W. Boldt

"To Abram Dück"

To: Peter Cornies
"Dear Cousin, Received your card and thank you very much. I am well and wish you the same. Furlough not before December 4, but then completed my service. Greeting to parents, siblings, and yourself.
Your Johann Fast"
June 14, 1915

Abram Dück, centre, Jacob Tiessen seated right

City Hall on Nikolaiev Street in Baku

Magnificent Alexandrovskii Park in Batum

High School for Ladies in Batum

The festive Marinskij Street in Batum

Abram Dück

From Jacob Tiessen's Collection

Johann Enns, standing
Peter Cornies in carriage

Peter Cornies   Johann Enns

A group photo

Planting trees

Entrance to a forestry camp

Lunch break

Typical poses of that period

Rolling countryside

Roasting fresh game

David Thiessen

Photos above and below: Close up views of the ruins near Ksanka

"Dear Brother,
Wishing you the best of health, that I, thanks and praise to God am also enjoying again. By the time you receive this card it will be Easter. Wishing all of you there a happy and blessed holiday. If I am not mistaken you will recognize one person on this picture. If you do, please write which one it is. If you should meet Peter Neufeld or Jakob Klassen, greet them. With heartfelt greetings, I remain your brother Johann."

Johann Tiessen

A beautiful mountain scene

From Peter Sawatzky's Collection

"Franz Sawatsky in his winter uniform at the Forstei.*"*

Franz Sawatzky

Dietrich Sawatzky

Jakob Warkentin

Gerhard Loewen, Johann Krahn, Johann Warkentin

Heinrich Janzen

"I am sending you a photograph of Uncle Hein. Here you can see how thin he already is. Now he is very ill and will weigh even less. He writes that in 5 days he lost [?] pounds, but will probably gain it back. The best would be if he could get well again. I received a letter from him yesterday and it says that he is still bedridden and won't be getting up soon. You can imagine how bad I feel. In closing, give my regards to Peter and Marichen and Hein and Mariechen and the others."

Johann Franz Sawatzky

Peter Sawatzky

Isaak H. Funk

"My comrade Peter Schroeder in the Crimea."

Molotschansk Gouv. of Taurien, Nov. 18, 1917

Peter Schroeder in the Crimea

"To the memory of your friend Jakob Wiens. Station Vorosivka, Voronezh Gub, P.O. Mushni Novoselsk"
Jakob Wiens

Jakob Wiens, Jakob Wiens at the South Coast forestry camp near Yalta

A Mr. Plett died at the South Coast forestry camp in 1916. He was the son of a bishop from the Molotschna.

Peter J. Sawatsky at the South Coast forestry camp, Crimea

"At the South Coast forestry camp near Yalta. Searching for a plot to plant a vegetable garden. From left to right: watchman Johann Giesbrecht, Johann Dyck and P.J. Sawatsky."

In search of a vegetable plot near Yalta, Crimea

Celebrations in honour of the Tsar

"Celebrating the coronation of the Tsar."

Marching to work

"Year 1911, August 14."

From Peter Warkentin's Collection

Johann Warkentin

Abram Dück is at the top left corner of this photograph and Jacob Tiessen on the far right.

Dietrich Sawatzky is at the top left of this photograph.

The fifth person from the right, in a reclining position, is Dietrich Sawatzky.

Epilogue

Tbilisi, Georgia
2002

By Jacob Dick

While writing and researching this book, I received a letter from the Canadian Executive Services Organization (CESO) asking if I had any interest in helping a small business in Tbilisi, capital of the Republic of Georgia. I could hardly believe what I had read. After contemplating the idea for several days, I concluded I would somehow make it work. The thought of going to Tbilisi to look for traces of my father's experiences was unbelievable.

On January 25, 2002, at one A.M., we were welcomed to Tbilisi by our excellent country representative, Asmat Abesadze. On the next day, Asmat bought us a city map in English, and walked with us through the old city and back to our hotel. We were then quite well oriented, and walked many miles observing the interesting city. The next day, Asmat took us to the Ethnographic Museum for a quick history lesson on Georgia, which is five thousand years old.

On the morning of Monday, January 28, we met our sponsor, MTEBI Ltd. MTEBI is a small company that was requesting help in their plan to export water to the Arabic countries. Early in our discussions with our hosts, I related my interest in Tbilisi and told them my father had been here with the Russian medical services during World War I. Our translator and guide, Levan Mchedlishvili, looked up at me for several seconds and asked me, "What was your fathers' name?" I said, "Abram." "Oh," replied Levan. He went on to say that his grandfather had been severely wounded on the Turkish Front during World War I. He was wonderfully cared for on a medical train, particularly by a German doctor named Ludwig, who he thought may have been my father. He also said that his grandfather named his father "Ludwig" after this doctor. We looked at each other in awe, considering the possibility that he had been on one of my father's trains.

Whenever my hosts asked me if there was something special I wanted to see, I always looked for locations relevant to my father and his colleagues. I told them I was interested in seeing the train station but, upon seeing it, was disappointed to see that the communists had rebuilt it.

My colleagues took me to the existing military hospital that, like the railway station, was a very large newer building, characteristic of

The old Tbilisi railway station.

The new Tbilisi railway station.

communist-era architecture. The hospital was still connected to the Russian military, and I was not allowed to take a picture. Nearby was an older building that looked to be of the World War I era.

On another occasion, I showed my colleagues a picture of a bridge from my father's collection of photographs. My colleagues identified it and told me where to find it, which I managed, after a rather long walk.

The old hospital in Tbilisi, as seen today.

The Voronzov Bridge and Mount David

The Voronzov Bridge at the time of publication

One day I showed my colleagues a picture of an old ruin with the name "Ksanka" written on the back and asked them if they knew where it was. After some time, they identified it as the Ksani ruin, and took me on a two-and-a-half-hour trip to see it the next day.

An early photograph of the Ksani ruins.

The Ksani ruins at the time of publication

The Ksani citadel dates back to the tenth or eleventh century. There is a legend associated with this ruin. The King was in this citadel when a governor and his supporters surrounded the fort with the intention of keeping the fort under siege until the king and his people starved to death or surrendered. However, the people in the fort managed to dig a tunnel to the nearby river, unbeknown to the governor. The fort was under siege all fall and winter, and in the spring the captives threw fresh fish over the walls of the fort. The governor then realized that he had been out-manoeuvred and retreated.

We spent one lovely, sunny afternoon touring the Botanical Gardens and the ruins of the Narikala Fortress nearby. The Botanical Gardens are over two hundred years old, and though sadly neglected now, there is much evidence of a glorious past. The gardens are set in a ravine beside the ruins, with many stairs, paths and walkways throughout. It is a beautiful setting.

One afternoon, my wife and I decided to visit the Tbilisi Museum, which was another pleasant surprise. In one section of the building, an area is dedicated to the Medical Services of World War I. It includes a display of photographs with a picture which also appears in my father's collection.

Through sign language and persistence with the Security Staff, I was eventually put in contact with an English-speaking member of the museum staff, Lika Mamasashvili. She was much interested in the copies of photographs I had with me and wanted to hear the stories behind the pictures. I expressed interest in an old railway photograph, and Lika was willing to send it to me by e-mail. She was true to her word, and three photographs arrived by e-mail the day after we arrived home. I promised to send her copies of some of my photographs, and we continue to exchange photographs from each other's collections.

I told my colleagues of my frustration with trying to match the names of cities mentioned in the story to those on current maps, and fortunately, they were able to help me. The elder colleague, Mr. Yura

Medical workers enjoying the botanical gardens of Tbilisi, Georgia.

A current picture of the botanical gardens in Tbilisi.

Bakhtadze, had an old atlas on his bookshelf dated 1914, on which all the old names were still used.

My most memorable experiences in Tbilisi were the people's reactions to our interest in their nation and its history. I will always remember their surprised and joyful responses.

— *Jacob Dick*

Made in United States
North Haven, CT
10 November 2026

22088272R00095